Novel Sensors and Sensing

Series in Sensors

Senior Series Editor: **B E Jones**
Series Co-Editor: **W D Spillman, Jr**

Other titles in the series

Forthcoming titles in the series

Series in Sensors

Novel Sensors and Sensing

R G Jackson

IoP

Institute of Physics Publishing
Bristol and Philadelphia

British Library Cataloguing-in-Publication Data

A catalogue record for this book is available from the British Library.

ISBN 0 7503 0989 X

Library of Congress Cataloging-in-Publication Data are available

Commissioning Editor: Tom Spicer
Production Editor: Simon Laurenson
Production Control: Leah Fielding
Cover Design: Victoria Le Billon
Marketing: Nicola Newey

Published by Institute of Physics Publishing, wholly owned by The Institute of Physics, London

Institute of Physics Publishing, Dirac House, Temple Back, Bristol BS1 6BE, UK

US Office: Institute of Physics Publishing, The Public Ledger Building, Suite 929, 150 South Independence Mall West, Philadelphia, PA 19106, USA

Typeset by Academic + Technical, Bristol
Printed in the UK by MPG Books Ltd, Bodmin, Cornwall

Contents

Preface

During the past twenty years there has been a tremendous growth in the research and development of sensors and associated signal processing systems. This book attempts to provide an insight into the generic operating principles of these novel developments. It does not cover the more traditional sensors such as metallic strain gauges, process industry thermocouples and orifice plate flowmeters. There are a number of good texts available that thoroughly cover such devices.

Some of the sensors described have not become commercially available yet and may not. However, it was felt that the principles of engineering physics involved warranted their mention. Often, commercial pressures change making uneconomic devices suddenly more attractive. This can also be precipitated by the development of new materials or manufacturing technology.

A book of this nature is bound to be somewhat idiosyncratic, covering the experience and interests of the author. It has been distilled from undergraduate lectures given by the author to students having completed a more traditional instrumentation module. Consequently, the book should be of interest to undergraduates of physical science/engineering and those practitioners in the field of instrumentation who wish to know the principles of this new generation of sensors and what might become commonly available in the near future.

Research papers are included among the references cited at the end of each chapter. These are not meant to show the latest developments in the field, since any such attempt would become out of date before publication. Rather, they are an indication of the historical context of the work.

Chapter 1

The instrumentation system

1.1 Introduction

Whether mankind is simply another part of the animal world or is somehow unique is a question that has occupied the minds of philosophers for generations. The human attributes of self-awareness, empathy, morality and what constitutes consciousness and creativity have become part of the current debate surrounding the development of machine intelligence. What is beyond question is the insatiable thirst mankind has to discover more about the individual self and the universe around. Observations for scientific purpose have been made since the beginnings of history and instruments to facilitate this have existed for millennia. However, during the latter half of the 20th century the growth in instrumentation technology, and in conjunction control theory, has been phenomenal.

For a nation's instrumentation industry to remain at the fore in world markets sensor development is essential. In addition to the general technological development of traditional sensors, during the past twenty years a number of specific fields have emerged centred on novel sensing techniques and materials. The global market for all sensor types is estimated at €15–30 billion per annum. The current growth in sales of traditional sensors is around 4% per annum compared with that of 10% per annum for novel sensors. (Source: UK Foresight Programme.)

Initially, the Japanese were the leaders at commercially exploiting academic innovations. However, the established industrial countries of the west are now firmly committed, both at governmental as well as company level, to strategic programmes for sensor development. Key fields are regularly identified by funding agencies in order to stimulate research and to assist in commercial exploitation.

The benefits of improvements made in sensors will enhance the quality and performance of industry and other fields. For example, it has been said that advances in process automation are presently limited to the quality of

1

the instrumentation and in particular the plant–instrument interface. In medicine, novel developments occurring in diagnostic instrumentation will reduce response time and costs, and increase throughput. Non-invasive sensors and invasive microsensors can reduce patient trauma. In the field of transportation, modern vehicles contain large numbers of sensors for safety, engine efficiency and global location systems. Medical, environmental and automotive applications have the largest potential market growth.

In any new device it is important that the specification or transfer characteristic should be well defined and stable. This is often the most difficult problem to resolve when going from laboratory prototype to commercial product. Many novel microsensor-prototypes have been reported measuring, for example, position, flow, humidity, acceleration, liquid level, ionic concentration, temperature, pressure and dissolved oxygen. However, the percentage of devices that has become commercially available is relatively small. Nevertheless, many of those that have enjoy large volume production which is encouraging the general growth of the novel sensor industry.

It is important to realize that to be commercially successful a device must be able to meet the cost constraints of the market for which it is

Table 1.1. Acceptable sensor parameters in some common fields.

	Domestic	Industrial	Medical	Automotive	Environmental
Acceptable cost (£)	1	500	Disposable 10, non-disposable 250	5	Disposable 10, non-disposable 250
Acceptable error (%)	10.0	0.1–5.0	0.1–2.0	1.0–5.0	1.0–10.0
Mean time between failures (hours)	10^3 during 10 years of intermittent use.	10^5 during 10 years of continuous use.	Disposable 150 during 1 week of continuous use, non-disposable 5×10^3 during 5 years of intermittent use.	1.5×10^4 during 10 years of intermittent use.	Disposable 150 during 1 week of continuous use, non-disposable 10^4 during 10 years of intermittent use.
Temperature range (°C)	−20 to 400	−200 to 1500	Disposable 10 to 50, non-disposable 0 to 250	−30 to 400	−30 to 100
Response time (secs)	1	0.1–10	1	10^{-3}–1	100

intended. Also, some fields of application require a higher specification than do others. Table 1.1 indicates acceptable costs and other desired parameters in some common fields.

Summarizing, it is important that for any newly developed sensor to become commercially successful it must not only be effectively 'ruggedized', when going from the laboratory prototype to marketed product, but also fit within the constraints of the application field.

1.2 The philosophy of measurement

Lord Kelvin is reputed to have said,

> 'When you can measure what you are speaking about and express it in numbers, you know something about it; but when you cannot measure it, when you cannot express it in numbers, your knowledge is of a meagre and unsatisfactory kind.'

Experimental observations are the basis on which our understanding of the universe around us develops. In science and engineering all hypotheses and theories before they develop into rules and laws have to be rigorously examined to discover their validity and its extent. Universal constants and parameters relating to materials and structures and other artefacts all have to be painstakingly examined to produce the most reliable data possible. A good example is the measurement of the speed of light, which has taxed some of the cleverest experimenters for centuries and for one man in particular, A A Michelson (1852–1931), took up a large proportion of his life earning him the Nobel Prize in 1907.

The philosophy of measurement is that of striving to reach some notional real or true value. The more effort made the greater the precision of the measurement and the smaller the uncertainty.

1.2.1 Random errors

Whenever the measurement of a parameter or variable is made an error will occur. The measured value or *measurand* will have an associated uncertainty due to the natural fluctuation of real physical quantities. Gauss postulated that such variations are as equally likely to be above the mean as below it. This *random* error may not be noticed if the measurement system is insensitive, and may not be of any concern in some applications. However, at the microscopic scale nature is never still and highly sensitive instruments will reveal such fluctuations. The instrumentation system is part of the physical world and will itself introduce an observable randomness to any measurements taken. All such random errors have to be treated statistically.

The best estimate of the true value X of some variable or parameter x is the mean $\bar{x}$. In the usual way,

$$\bar{x} = \frac{\sum_{i=1}^{N} x_i}{N} \tag{1.1}$$

from N observations. It is expected that for large N, $\bar{x} \to X$. Incidentally, when calculating the mean it is argued that the numerical value obtained from a large number of observations can be rounded to an extra significant figure more than that of the precision of the equipment used.

A measure of the spread or dispersion of the results around the mean is very useful as an indication of the repeatability of the measurement. The deviation of a single result from the mean is $x_i - \bar{x}$. Because the deviations can be positive or negative the sum of the deviations is likely to be zero for a large set of values. Consequently, the square of the deviations is used. The mean of the squared deviations is called the *variance*. To obtain a measure of the dispersion in the same units as the original measurements the square root of the variance must be taken. This is the *standard deviation*. It is simply determined using

$$s = \sqrt{\frac{\sum_{i=1}^{N} (x_i - \bar{x})^2}{N}}. \tag{1.2}$$

The standard deviation is also known as the *standard error*. It is a measure of the randomness or fluctuations in the results obtained. It is independent of N, the number of observations made. As can be seen in equation (1.2) making many, repeated observations does not reduce the value of s but simply gives a greater confidence in its estimate.

Intuitively it is felt that taking more measurements ought to improve the quality of the mean in some quantifiable way, 'more must be better'. To provide this extra quantifier imagine that instead of one set of N results the measurement set was repeated p times under identical conditions, each set having N results. Each of the p sets of results could produce its own mean $\bar{x}$ and standard deviation as before. But now these p means could be averaged to give the mean of the means $\bar{\bar{x}}$:

$$\bar{\bar{x}} = \frac{\sum_{k=1}^{p} \bar{x}_k}{p} = \frac{1}{p} \sum_{k=1}^{p} \left(\frac{1}{N} \sum_{i=1}^{N} x_i \right). \tag{1.3}$$

This does not look anything of significance. After all, it is the same result as would have been obtained having made pN observations initially:

$$\bar{\bar{x}} = \bar{x} = \frac{\sum_{i=1}^{pN} x_i}{pN}. \tag{1.4}$$

The most significant and useful result is obtained from the standard deviation of the p means. This standard deviation s_m is a measure of the spread of the means and indicates the precision of the value of the mean of the means. In principle this could be determined using

$$s_{\mathrm{m}} = \sqrt{\frac{\sum_{k=1}^{p} (\bar{x}_k - \bar{\bar{x}})^2}{p}}.$$ (1.5)

In practice equation (1.5) is not used. It can be shown (see Squires 1968) that

$$s_{\mathrm{m}} \cong \frac{s}{\sqrt{N-1}} = \pm \sqrt{\frac{\sum_{i=1}^{N} (x_i - \bar{x})^2}{N(N-1)}}.$$ (1.6)

So for large N,

$$s_{\mathrm{m}} \rightarrow \pm \sqrt{\frac{\sum_{i=1}^{N} (x_i - \bar{x})^2}{N^2}}.$$ (1.7)

As equation (1.7) indicates, the estimate of s_{m} can be seen to improve by increasing the number of observations made. 'More is better' is correct here. However, this rule suffers from diminishing returns. Large values of N may not be worth the effort. To appreciate the different meanings of s and s_{m} the following examples may be helpful.

Standard error of the measurements (s)

Consider the manufacture of $10\,\mathrm{k}\Omega$ resistors on a production line. Each day 100 samples could be taken from the production run and the value of s determined. This denotes the spread of the results about the mean and indicates to the production manager whether the batch is within the tolerance limits specified on the packets, e.g. $\pm 10\%$. The probability of a single resistor being outside of the $\pm 10\%$ is related to s. In section 6.3.2 this is described further within the topic on probability distributions. For a value to be more than three standard errors from the mean is statistically unlikely at about 2.7 in 10^3. So, if s is arranged to be about 3%, by using good production technology, then most resistors produced will be within 10% of the marketed value of $10\,\mathrm{k}\Omega$.

This '3s' concept can also be applied to identify an erroneous result. If a reading from a set of identical measurements is more than three standard errors from the mean of the set it can be considered dubious and be discarded.

Standard error in the mean (s_{m})

In experiments where some parameter is to be determined, such as the speed of light, the value of s_{m} is quoted. This provides an estimate of the precision of the published value and incorporates the effort made in producing the repeated sets of measurements. There have been occasions when researchers

Table 1.2. Combination of standard errors.

The function $f(X, Y, Z)$	Combination of standard errors
$Q = X + Y + Z$ or $Q = X + Y - Z$	$s_Q^2 = s_X^2 + s_Y^2 + s_Z^2$
$Q = XYZ$ or $Q = \dfrac{XZ}{Y}$	$\left(\dfrac{s_Q}{Q}\right)^2 = \left(\dfrac{s_X}{X}\right)^2 + \left(\dfrac{s_Y}{Y}\right)^2 + \left(\dfrac{s_Z}{Z}\right)^2$
$Q = X^n$	$\dfrac{s_Q}{Q} = n\dfrac{s_X}{X}$
$Q = \ln(X)$	$s_Q = \dfrac{s_X}{X}$
$Q = \exp(X)$	$\dfrac{s_Q}{Q} = s_X$

have repeated experiments and obtained results outside of the standard error quoted originally. This has led to the discovery of systematic errors.

Combining standard deviations

If a parameter or variable Q is to be determined from measurements of two or more parameters or variables X, Y, Z each having their own standard deviations s_X, s_Y, s_Z, it will be necessary to determine the standard deviation of Q, s_Q, by combining these contributing standard errors. It can be shown that if Q is a function $f(X, Y, Z)$ then

$$s_Q^2 = \left(\frac{\partial f}{\partial x}\right)^2 s_X^2 + \left(\frac{\partial f}{\partial y}\right)^2 s_Y^2 + \left(\frac{\partial f}{\partial z}\right)^2 s_Z^2. \tag{1.8}$$

Table 1.2 shows how to combine the standard errors for some common functions.

1.2.2 Systematic errors

A further error may occur due to poor calibration or poor performance in the instrumentation system. Because such errors are inherent in the system and are transmitted through to the measured value they are termed *systematic* errors. They cannot be overcome by the statistical analysis of multiple readings but are minimized by diligence in calibration and system design.

1.2.3 Environmental disturbances

The engineering environment within which the instrumentation system is integrated can affect the input–output relationship of the sensor or of the instrumentation system as a whole. The engineering environment includes

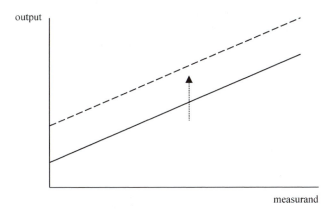

output

measurand

Figure 1.1. The interfering effect.

variables such as the ambient temperature and humidity and power supply behaviour. There are two common ways in which the environment can affect the response and are described below for an idealized, linear sensor or system.

Zero-shift or the interfering effect

This disturbance has the effect of changing the intercept value of the straight-line graph, shown in figure 1.1. The interfering input is the change in or deviation of some environmental parameter. This input couples into the system via a sensitivity constant.

For example, if a drift in the ambient temperature of 5 °C caused a pressure measurement reading to change by 150 Pa then the system would have an environmental sensitivity of 30 Pa K^{-1}.

Variation in sensitivity or the modifying effect

This environmental disturbance modifies the value of the system sensitivity, the gradient of the graph (figure 1.2). An example of this effect could be of the float-type liquid level gauge as commonly used in automobiles. A potentiometer fastened to the float provides a variable resistance related to the level of the liquid, which in turn alters the current through the milliammeter petrol gauge. If the supply voltage alters then so does the reading. Typically, when the alternator and regulator are operating, an automobile power supply is held steady at 14 V. However, when the engine is stopped the power supply system is reduced to that of the 12 V battery. An environmental sensitivity constant could be determined in gallons mA^{-1}/V.

A simple way to monitor whether there has been some environmental disturbance is to repeat the first measurement taken after the completion

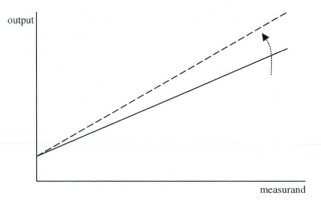

output

measurand

Figure 1.2. The modifying effect.

of a set of readings. More sophisticated methods can be devised. For example, for a linear system denoted by $y = mx + c$ the plotting of $y - mx$ versus x should yield a fixed value if c is constant (no interfering effect). Similarly, plotting $(y - c)/x$ versus x should give a constant if m is fixed (no modifying effect).

1.2.4   System design and the contribution of errors

It was shown in section 1.2.1 how the standard errors of a number of random variables can be combined to give the standard error in some derived quantity Q. An appreciation of the way errors combine is also useful when an experiment or system is being designed. It can be instructive to see how errors in the component measurands of the quantity to be determined and any additional system element errors contribute to the final result. In this way more care can be taken over those components that contribute most to the global error.

As previously, suppose a quantity Q is to be determined from observations of one or more variables. In addition, other elements in the instrumentation system such as a filter or an amplifier may also introduce errors. Again, Q is some function $f(X, Y, Z)$ where X and Y might be the measurands and Z the gain of an amplifier. The values of X, Y and Z have associated uncertainties δX, δY, δZ which combine to produce an uncertainty δQ in Q. The relationship between δQ and δX, δY, δZ can be found using the partial differential equation

$$\delta Q = \frac{\partial f}{\partial X} \delta X + \frac{\partial f}{\partial Y} \delta Y + \frac{\partial f}{\partial Z} \delta Z. \tag{1.9}$$

Table 1.3 gives examples of some common functions.

Table 1.3. Simple combination of errors.

The function $f(X, Y, Z)$	Combination of standard errors
$Q = X + Y + Z$ or $Q = X + Y - Z$	$\delta Q = \delta X + \delta Y + \delta Z$
$Q = XYZ$ or $Q = \dfrac{XZ}{Y}$	$\dfrac{\delta Q}{Q} = \dfrac{\delta X}{X} + \dfrac{\delta Y}{Y} + \dfrac{\delta Z}{Z}$
$Q = X^n$	$\dfrac{\delta Q}{Q} = n\dfrac{\delta X}{X}$
$Q = \ln(X)$	$\delta Q = \dfrac{\delta X}{X}$
$Q = \exp(X)$	$\dfrac{\delta Q}{Q} = \delta X$

Addition/subtraction

Unless there is some *a priori* knowledge about the system behaviour, the worst case is assumed and all the *absolute* errors are added. The error estimates must all be in the same units and δQ will also have these units.

Multiplication/division

Here, again assuming the worst case, all the *relative* errors are added. They have no units and can be described as fractional errors or multiplied by 100% to give percentage errors. The relative error is also used with the logarithmic function.

Simple determination of the global error

The relationships in table 1.3 or others derived from equation (1.9) can also be used to make a simple estimate of the error in a derived quantity, whether determined from a number of measurands, or as the output from an instrumentation system. These contributory errors might arise from

- a rough guess at the random errors,
- estimated systematic errors,
- calibration limits,
- component tolerances,
- estimated likely environmental disturbance,
- the resolution limits,
- hysteresis or 'dead band' effects,
- estimated nonlinearities,
- digitization errors.

In the paragraph preceding equation (1.9) the word *uncertainty* was used when describing δX, δY, δZ. When dealing with the quantities listed above this is a preferred expression to *error* since it is all encompassing. Items such as calibration limits and tolerances are then included which, strictly speaking, are not errors.

WORKED EXAMPLE

The density of a material is to be determined from the mass and volume of a wire sample and a simple method used to estimate the uncertainty in the answer obtained. The density is derived from

$$\rho = \frac{m}{\pi(d/2)^2 h}.$$

From equation (1.9) it can be seen that

$$\frac{\delta\rho}{\rho} = \frac{\delta m}{m} + 2\frac{\delta d}{d} + \frac{\delta h}{h}$$

revealing that the fractional uncertainty in the diameter d is the most sensitive term in the determination of ρ. Greater care must therefore be taken with this measurement.

The mass of the specimen is found to be 1.25×10^{-3} kg and the balance used is calibrated to ± 1 mg. The length is measured as 0.1845 m using a rule readable to the nearest 0.5 mm. The thickness is measured ten times at various points and orientations along the length using a micrometer capable of measuring to $\pm 10^{-5}$ m and is found to be 0.984 mm with a standard error in the mean of ± 0.024 mm. The density is determined as

$$\rho = \frac{1.25 \times 10^{-3}}{\pi\left(\dfrac{9.84 \times 10^{-4}}{2}\right)^2 \times 0.1845} = 8.909 \times 10^3 \text{ kg m}^{-3}.$$

The fractional uncertainty is determined,

$$\frac{\delta\rho}{\rho} = \frac{10^{-6}}{1.25 \times 10^{-3}} + 2\frac{0.024}{0.984} + \frac{0.5}{184.5}$$

$$= 8 \times 10^{-4} + 0.04878 + 2.71 \times 10^{-3}$$

$$= 0.05229.$$

The uncertainty in ρ is

$$\delta\rho = \rho \times 0.05229 = 465.9 \text{ kg m}^{-3}.$$

It is usual to express the result in the form

$$\rho = 8.909 \pm 0.4659 \times 10^3 \text{ kg m}^{-3}.$$

The question of the number of significant figures quoted needs to be considered. A common sense approach must be adopted. In the example above, considering that the uncertainty is about 5% (mostly from the diameter error), it would seem meaningless to quote the final value for ρ to more than 2 significant figures

$$\rho = 8.9 \pm 0.5 \times 10^3 \, \text{kg m}^{-3}.$$

The uncertainty or standard error is normally quoted to a single digit or to two digits if the digit is 1 or 2.

Finally, it is important to realize that this simple method of combining errors does not have the statistical rigour of that used for combining standard errors achieved by equation (1.8). This is because the uncertainties here do not belong to probability distributions but are bands of 'vagueness'. However, on occasion it may be possible to consider the uncertainty as inherently part of a gaussian distribution. The uncertainty could then be interpreted as a standard error and equation (1.8) invoked to combine it with other standard errors. The advantage of this is that the standard error produced gives a precise, statistical weighting as to the reliability of the quantity. In section 6.3.2 the method of calculating such probabilities is illustrated.

1.3   The general instrumentation system

The instrumentation system can be divided conveniently into the following general components. Not all systems contain all components.

input value G_1 Sensor: This is in good contact with the variable to be measured and some physical property of the sensor is modified by the variable. On occasions the sensor may be composed of two elements termed primary and secondary.

p

G_2 Signal conditioning: The output of the sensor may not be in a convenient format. The signal conditioning circuit will most commonly be used to provide an output of a few volts.

q

G_3 Signal processing: The few volts of analogue data may require filtering, linearizing, converting to digital format or preparing for transmission if a telemetry link is required.

Telemetry link: May be wire, pneumatic, microwave or optical fibre and use AM, FM or pulse code.

r

measured value G_4 Data display: This may be analogue or digital as appropriate for the application. The ergonomics of data displays can be of significance. Table 1.4 lists some of the relative characteristics of each type.

Table 1.4. Analogue versus digital displays.

	Advantages	Disadvantages
Analogue	Continuous scale. Faster recognition by user	Less easy to read small divisions. Easier to misread
Digital	Cheaper. Faster response	More stressful in continuous use. Wrong value shown if a segment fails

To conclude this section it was felt that a comment on the terminology in current use concerning sensors would be appropriate. The word *transducer*, strictly defined, is a device that converts energy from one form to another. A microphone could be described as a transducer but so could a loudspeaker. A sensor might be a transducer but a transducer may not be a sensor. However, a recent advertisement for a *pressure transducer*, on inspection, was found to be a diaphragm with strain gauge and associated signal conditioning. A power supply was input on two wires and the output on a further two. This, clearly, is not a transducer in the strict sense.

A further example of misleading terminology is the use in the process industries of the word transmitter. A *pressure transmitter*, for example, is a device that measures the required pressure and outputs the information on a wire pair usually in the standard industrial range of 4–20 mA. Such *transmitters* are sensor, signal conditioning and signal processing.

It is of paramount importance in science and engineering that terms are clearly defined and upheld. History is littered with accidents caused by misunderstandings between engineers. The modern trend is to use the word *sensor* for an input device and *actuator* for the output device attempting to cause some change to a system or plant. The word transducer is best avoided. Sensors can be subdivided into *active*, when energy conversion occurs producing an electrical signal, and *passive* when a power supply is required to obtain an electrical output. For example, a thermocouple is an active sensor whereas a resistance thermometer a passive sensor. This distinction can be significant when dealing with explosive environments where the energy available within the hazardous area must be strictly controlled.

1.4   The overall transfer function

The output of the instrumentation system is a product of the input variable and the various elements within the system. In analysing the overall system performance it can be useful to mathematically model each element. Finkelstein and Watts (1971) proposed this systems approach to instrumentation.

The transfer function for each element is

$$\frac{\text{output}}{\text{input}}.$$

For example, if the sensor were a resistance temperature element then

$$G_1 = \frac{p}{\text{input value}} \Omega\,{}^\circ\text{C}^{-1}. \qquad (1.10)$$

Individual elements may contain feedback loops but this is incorporated into the value of G. The overall transfer function is the cascade of the individual transfer functions

$$\frac{\text{measured value}}{\text{input value}} = G_1 G_2 G_3 G_4 G_5. \qquad (1.11)$$

The overall transfer function is the sensitivity of the system. Clearly then

$$\text{measured value} = \text{input value} \times G_1 G_2 G_3 G_4 G_5. \qquad (1.12)$$

However, this will not be the case if an element has a zero or an off-set value. For example, the standard platinum resistance sensor has a resistance of $100.0\,\Omega$ at $0\,{}^\circ\text{C}$. To overcome this the signal-conditioning element is made to output zero for an input of $100.0\,\Omega$. It is good practice generally to have off-set or zero adjust facilities early in the system. Failure to do this could result in large signals later in the system causing saturation of an amplifier or inefficient use of the input span of an analogue to digital converter.

Similarly, the expression (1.12) above takes no account of inter-element loading. Again it is good practice to construct elements with low output impedance and high input impedance to minimize loading effects. The availability of high performance operational amplifiers for use as buffers makes this requirement easy to achieve. Inter-element loading is described further in section 1.6.

1.5   Dynamic response of the sensor

The modelling of the sensor itself can be more involved than equation (1.10) suggests. For example the warming or cooling of a temperature sensor can be modelled by a first-order, linear, differential equation. This means that because it is not an algebraic expression the transfer function cannot be simply set down. Generally, elements that involve energy storage and transfer are described by models involving time-dependent equations. Consequently, if a variable is changing, the output measured value will have a *dynamic* error. To be able to use the cascade concept above for the whole system it is convenient to Laplace transform the element equations. This has the effect of turning an expression involving calculus into one involving algebra. The example below illustrates the dynamic response of a temperature sensor.

Transfer function of a temperature sensor

At the start of the measurement the sensor will have a different temperature from that of the fluid in which it is immersed. There must be an energy transfer between fluid and sensor until they are both at the same temperature. At the start

time $t = 0$,   sensor temperature $= \theta_0$,   fluid temperature $= \theta_f$.

During the heat transfer process the sensor temperature will be changing continuously and will have some value $\theta_s(t)$. To cause the temperature of some material to rise a quantity of heat Q must be supplied,

$$Q = mc(\theta_s - \theta_0) \tag{1.13}$$

where m is the mass of the material and c its specific heat capacity. Assuming that the fluid is hotter than the sensor, the rate of gain of heat by the sensor is

$$\frac{dQ}{dt} = mc\frac{d(\theta_s - \theta_0)}{dt}. \tag{1.14}$$

The heat transfer can be by either convection, conduction or radiation. In this situation the dominant process will be convection. The rate of heat loss by the liquid is determined using the convection equation

$$\frac{dQ}{dt} = UA(\theta_f - \theta_s) \tag{1.15}$$

where A is the surface area and U is a geometry factor.

When using Laplace transforms it is more convenient to have zero starting conditions. This means that all variables in the equation should be zero at the start. To arrange for this a substitution is made,

$$\Delta\theta \equiv \theta_s - \theta_0. \tag{1.16}$$

Equation (1.14) now becomes

$$\frac{dQ}{dt} = mc\frac{d\Delta\theta}{dt} \tag{1.17}$$

and equation (1.15) becomes

$$\frac{dQ}{dt} = UA[\theta_f - (\Delta\theta + \theta_0)]. \tag{1.18}$$

Equation (1.18) can be written more conveniently using the substitution

$$\Delta\theta_f = \theta_f - \theta_0 \tag{1.19}$$

giving

$$\frac{dQ}{dt} = UA(\Delta\theta_f - \Delta\theta). \tag{1.20}$$

Now,

> the heat gained by the sensor = the heat lost by the fluid.

So, combining equations (1.17) and (1.20) gives

$$mc\frac{d\Delta\theta}{dt} = UA(\Delta\theta_f - \Delta\theta) \tag{1.21}$$

and re-arranging gives

$$\frac{mc}{UA}\frac{d\Delta\theta}{dt} + \Delta\theta = \Delta\theta_f. \tag{1.22}$$

Science and engineering equations must be homogenous in their units. The coefficient in front of the differential term must therefore have the dimensions of time. A time constant term τ can be introduced where

$$\tau = \frac{mc}{UA}. \tag{1.23}$$

Substituting τ into equation (1.22) gives

$$\tau\frac{d\Delta\theta}{dt} + \Delta\theta = \Delta\theta_f. \tag{1.24}$$

This can now be conveniently transformed using Laplace tables to give

$$\tau s\Delta\bar{\theta}(s) + \Delta\bar{\theta}(s) = \Delta\bar{\theta}_f(s) \tag{1.25}$$

$$\Delta\bar{\theta}(s)(\tau s + 1) = \Delta\bar{\theta}_f(s) \tag{1.26}$$

$$\Delta\bar{\theta}(s) = \frac{\Delta\bar{\theta}_f(s)}{1 + s\tau}. \tag{1.27}$$

So the Laplace transformed transfer function of the sensor is

$$\frac{\Delta\bar{\theta}(s)}{\Delta\bar{\theta}_f(s)} = \frac{1}{1 + s\tau}. \tag{1.28}$$

Control engineers commonly use the systems approach in design and, conventionally, express transfer functions in Laplace transformed format. Upper case symbols are used to signify Laplace format so equation (1.28) would be given the symbol

$$G_1(s) = \frac{\Delta\bar{\theta}(s)}{\Delta\bar{\theta}_f(s)} = \frac{1}{1 + s\tau}. \tag{1.29}$$

1.6   The measurement system as a series of networks

Finkelstein and Watts (1971) proposed that elements in an instrumentation system could be described as two-port networks. Two-port network analysis is a technique whereby an electrical circuit, or part of a circuit, is treated as a generalized element having an input and an output port. The input/output variables are generally referred to as *across* (potential difference) and *through* (current) (Kraus 1991).

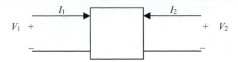

Figure 1.3. The two-port network.

Of the four variables two are chosen as independent and the others become the dependent. Depending on the choices made six *canonical* forms can be produced relating the input–output variables. A set of equations is derived, using a particular canon, which provides the generalized description. These equations may involve integral-differential relationships. Of the six canonical forms the impedance parameter equations are the most commonly used. For figure 1.3 these are

$$V_1 = z_{11}I_1 + z_{12}I_2 \tag{1.30}$$

and

$$V_2 = z_{21}I_1 + z_{22}I_2. \tag{1.31}$$

These can be conveniently expressed in matrix form as

$$\begin{bmatrix} V_1 \\ V_2 \end{bmatrix} = Z \begin{bmatrix} I_1 \\ I_2 \end{bmatrix} = \begin{bmatrix} z_{11} & z_{12} \\ z_{21} & z_{22} \end{bmatrix} \begin{bmatrix} I_1 \\ I_2 \end{bmatrix} \tag{1.32}$$

where Z is termed the impedance parameter matrix. Having described the elements in a system as sets of two-port networks with associated linear equations, matrix algebra or software tools such as Matlab® can be conveniently used to determine the overall system response.

Finkelstein and Watts suggested that the process to which the measured variable belongs could be treated as a one-port network. The process must have appropriately defined across (y) and through ($\dot{x}$) variables.

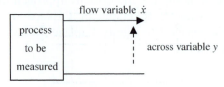

Figure 1.4. The process as a one-port network.

Table 1.5. Across and through variables for some common systems.

Variable	Electrical	Thermal	Fluid	Translational motion	Rotational motion
Across y	Potential difference (V)	Temperature difference (°C)	Pressure difference (Pa)	Force (N)	Torque (Nm)
Through $\dot{x}$	Current (A)	Thermal power (W)	Volume flow rate (m³ s⁻¹)	Velocity (m s⁻¹)	Angular velocity (rad s⁻¹)
x	Charge (C)	Heat energy (J)	Volume (m³)	Displacement (m)	Angular displacement (rad)
$\ddot{x}$	Rate of change of current (A s⁻¹)			Acceleration (m s⁻²)	Angular acceleration (rad s⁻²)
Impedance Z	Impedance (Ω)	Thermal resistance (°C W⁻¹)	Pressure/ volume flow rate	Force/ velocity damping constant (N s m⁻¹)	Torque/angular velocity damping constant (N m s rad⁻¹)

Bentley (1995) describes this approach in more detail. Some examples of across and through variables are shown in table 1.5.

The comparison of the charging of a capacitor and the heating of an object provides a simple example of the analogy that can be drawn between physical systems. For a simple RC network the output voltage V_s is developed from the fixed supply voltage V_f (figure 1.5). The time dependent differential equation is derived below.

For a capacitor,

$$CV_s = q. \tag{1.33}$$

Differentiating equation (1.33),

$$C\frac{\mathrm{d}V_s}{\mathrm{d}t} = i \tag{1.34}$$

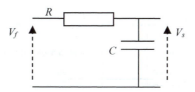

Figure 1.5. The analogy between a capacitor being charged and an object being heated.

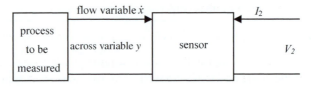

Figure 1.6. Process and sensor represented as one- and two-port networks.

$$C\frac{\mathrm{d}V_\mathrm{s}}{\mathrm{d}t} = \frac{V_\mathrm{f} - V_\mathrm{s}}{R} \tag{1.35}$$

$$RC\frac{\mathrm{d}V_\mathrm{s}}{\mathrm{d}t} = V_\mathrm{f} - V_\mathrm{s} \tag{1.36}$$

$$RC\frac{\mathrm{d}V_\mathrm{s}}{\mathrm{d}t} + V_\mathrm{s} = V_\mathrm{f}. \tag{1.37}$$

By comparing with equation (1.22) above it can be seen that

$$\frac{mc}{UA} = RC. \tag{1.38}$$

So, intuitively

$$C \equiv mc \qquad \text{thermal capacity}$$

$$R \equiv \frac{1}{UA} \qquad \text{thermal resistance.}$$

Concepts such as these are displayed in table 1.3.

If a process is being measured then a two-port network is added to represent the sensor (see figure 1.6). If, for example, the process is a hot tank whose temperature is to be measured by a resistance thermometer then the use of network analysis enables loading to be accounted for. The loading can occur in the usual way by the subsequent signal conditioning circuitry affecting the sensor. However, it can also occur by the sensor affecting the process it is measuring. In this example the former could be caused by the signal conditioning circuit heating the sensor element and the latter by the thermal capacity of the sensor element altering the temperature of the hot tank. The principles of impedance matching can be applied to the juxtaposition of process and sensor in order to get optimum response (figure 1.7). If the requirement is that maximum power is transferred from process to sensor

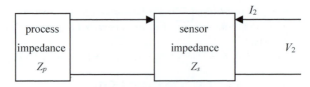

Figure 1.7. Impedance matching of the process and sensor.

then the usual condition applies that

$$Z_p = Z_s. \tag{1.39}$$

If, as is most commonly the case, the across variable generated by the process y_p should not be affected by the presence of the sensor then a different condition applies. Assuming that the flow variable is conserved around the system,

$$\text{the flow variable,} \quad \dot{x} = \frac{y_p}{Z_p + Z_s} \tag{1.40}$$

$$\text{the measured across variable,} \quad y_s = \dot{x}Z_s. \tag{1.41}$$

Eliminating $\dot{x}$ gives,

$$y_s = \frac{y_p Z_s}{Z_p + Z_s}. \tag{1.42}$$

So, for $y_s \cong y_p$, then $Z_s \gg Z_p$.

If the flow variable $\dot{x}$ is the desired measurand then the requirement would be that the sensor impedance should be as low as possible so that $\dot{x}$ is not restricted. For the process not to be affected it can be seen from equation (1.42) that the condition now is that $Z_p \gg Z_s$.

By the use of the impedance matrix technique it is possible to model the whole instrumentation system as an interconnecting group of elements driven by the process. In this way the dynamic response to a time-varying measurand can be determined before construction, making the design of complex systems more efficient.

Bibliography

This chapter has outlined the structure and general operation of an instrumentation system. The following chapters describe some of the more recent developments in sensors. Traditional sensors such as thermocouples, resistance thermometers and orifice-plate flowmeters have not been included since these are well covered in other texts. For readers wishing to know more of traditional sensors the following might be helpful.

Bentley J P 1995 *Principles of Measurement Systems* 3rd edition (Longman)
Considine D M (ed) 1999 *Process/Industrial Instruments and Controls Handbook* 5th edition (McGraw-Hill)
Johnson C D 1988 *Process Control Instrumentation Technology* 3rd edition (New York: Wiley)
Sensors: A Comprehensive Survey 1992 (Cambridge: UCH)
Usher M J 1985 *Sensors and Transducers* (Macmillan)

References

Bentley J P 1995 *Principles of Measurement Systems* 3rd edition (Longman)
Finkelstein L and Watts R D 1971 'Systems analysis of instruments' *Measurement and Control* **4** September
Kraus A D 1991 *Circuit Analysis* (West Publishing Company)
Squires G L 1968 *Practical Physics* (London: McGraw-Hill)

Exercises

1. A miniature camera has been developed with a radio telemetry link. It is cylindrical, 8 mm in diameter, 10 mm long and is battery powered. The probe costs £300 and the receiver system £2000 with a range of 10 m. Suggest some applications.

2. Suggest a suitable VDU system for the control panel in the engine room of a submarine.

3. The following measurements are made of the resistance R of a coil of wire,

Resistance R (Ω)	5.615	5.622	5.624	5.618	5.620	5.633	5.628	5.624	5.613

   Determine the best estimate of the resistance. Calculate the standard deviation of the measurements (s) and the standard deviation in the mean (s_m). Which standard deviation value should be quoted with R and why?

4. A flowmeter produces an output current which is proportional to the square of the flow velocity. Under steady flow conditions the readings of the output current are found to have a standard deviation of $\pm5\%$. What is the standard deviation of the flow velocity?

5. The time period of a simple pendulum is given by

$$T = 2\pi\sqrt{l/g}.$$

   A number of experiments are performed and it is found that the standard error in the mean of the time period T is $\pm0.7\%$ and of the length l is $\pm0.4\%$. Calculate the standard error in the mean of g.

6. If the errors given in question 5 were not standard errors but uncertainty estimates from calibration limits or similar what would the estimate of the overall uncertainty be then? Why is this estimate less useful statistically than the standard error produced in the more formal way of question 5?

7. A certain force sensor can be modelled as a spring, mass and damper whose dynamic equation is

$$F = m\frac{d^2x}{dt^2} + \lambda\frac{dx}{dt} + kx$$

where F is the applied force, m the effective mass, λ the damping constant and k the spring stiffness constant. Using Laplace transform tables and assuming zero initial starting conditions show that the transfer function $G(s)$ is

$$G(s) = \frac{x(s)}{y(s)} = \frac{1}{(1/\omega_n^2)s^2 + (2\xi/\omega_n)s + 1}$$

where the undamped natural frequency ω_n is $\sqrt{k/m}$ and the so-called damping ratio ξ is $\lambda/2\sqrt{km}$.

8. Finkelstein and Watts proposed that an instrumentation system could be described as a series of two-port networks:

   (i) Draw a block diagram of a general instrumentation system showing the various elements.
   (ii) In an electrical two-port network what are the *across* and *through* variables?
   (iii) In a thermal two-port network what are the across and through variables?
   (iv) A process to be measured can be described as a one-port network. What would be the across and through variables for a hot tank of liquid whose temperature is to be measured?

9. It is possible to use software packages such as the matrix based Matlab® to model an instrumentation system. The general two-port network can be described by the following block diagram.

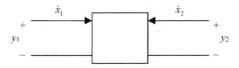

   (i) Write down the two impedance parameter equations.
   (ii) Re-write the impedance parameter equations in matrix format.

10. The temperature of a hot liquid of small volume is to be measured by a thermometer of thermal impedance Z_t.
   (i) What does it mean to say that the sensor might load the process?
   (ii) The block diagram for such a system can be represented by the following diagram

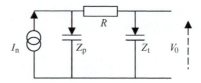

where I_n is the rate of supply of heat energy to the process of thermal impedance Z_p, R is the thermal resistance between the process and sensor and V_0 is the measured temperature. Show that

$$\frac{V_0}{I_n} = \frac{Z_t}{((R + Z_t)/Z_p) + 1}$$

(iii) Show that the Laplace transform of the equation above is

$$\frac{\bar{V}_0(s)}{\bar{I}_n(s)} = \frac{1}{C_p C_t R s^2 + (C_p + C_t)s}$$

where C_p and C_t are the thermal capacitances of the process and thermometer respectively.

(iv) Use the Matlab® toolbox Simulink® to model the transfer function in part (iii) above. (Hint: use the *Pulse* source with an appropriate height and width to deliver some heat Q.)

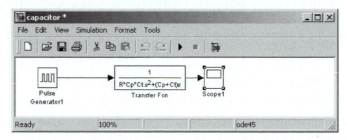

Chapter 2

Resonator sensors

2.1 Introduction

Resonator sensors are mechanical structures driven at their resonant frequency. It is arranged in the design of the sensor structure that the parameter to be measured alters the resonant frequency or the frequency spectrum or both. Unfortunately, the resonant frequency is generally a nonlinear function of the measurand and is subject to temperature drift. Also, mathematical modelling of all but the simplest structures is not easy. However, the dominant feature of resonator sensors that outweighs these disadvantages is their very high sensitivity. Having the output signal as a variable frequency can also be an advantage in environments subject to electromagnetic interference. But this does require more complex signal processing to recover the measured value.

Although the concept of resonator sensors is not new, with the advent of new materials and advances in modelling and fabrication techniques, smaller and more elegant designs are being developed. Sensors have been reported to measure mass, fluid flow, density, viscosity, rotation rate, force, pressure and acceleration. Microsensors using resonant techniques are in particular attracting a lot of interest because of the well-developed semiconductor fabrication techniques. Such sensors not only offer miniaturization, but also batch production, which lowers costs. Some of these devices are described in section 3.1.

Commonly used structures are wires, beams or cylinders. Because the output is a frequency, signal processing is straightforward and problems with noise and interference are significantly reduced. Also, since the device is being operated at resonance, power consumption is low. The main disadvantages are that the resonant frequency can be significantly altered by temperature fluctuations and that the output frequency is not linearly proportional to the measured parameter.

The oscillation process can be designed as either

- **harmonic**: energy is transferred from one stored form to another during a cycle; losses are made good by a power supply, or
- **relaxation**: only one form of energy storage exists; energy is regularly supplied in bursts and is dissipated by the oscillatory process between bursts.

The sensing part of the oscillatory structure may be at a point or small region (lumped) as in the case of the spring, mass and damper, or spread over the whole structure (distributed) as with the longitudinally vibrating cylinder.

This chapter describes some of the structures that have been produced in prototype form or as a commercial product during the past 25 years. They are not microsensors, although it is likely that some of the structures below will be adapted for use in miniature form as fabrication techniques develop.

2.2   The damped harmonic oscillator

Often, resonating systems can be described by a linear, second order, differential equation over some if not all of their range. The standard equation commonly used in engineering for any linear, second order, damped system is of the form

$$\frac{1}{\omega_n^2}\frac{d^2x}{dt^2} + \frac{2\xi}{\omega_n}\frac{dx}{dt} + x = y \tag{2.1}$$

where $y(t)$ is the forcing term, $x(t)$ is the resulting displacement, ω_n is the undamped natural frequency, and ξ is the damping ratio (if $\xi < 1$ the system response to an impulse will be oscillatory).

The format of equation (2.1) has become established because of the convenience of having ω_n and ξ as coefficients, which are easily identified. When comparison is made with the equation for some real system then the undamped natural frequency and whether oscillation will occur can quickly be determined. The simple spring, mass and damper system, for example, is modelled by

$$m\frac{d^2x}{dt^2} + \gamma\frac{dx}{dt} + kx = F \tag{2.2}$$

where m is the effective oscillating mass, γ is the damping constant, k is the spring stiffness and F is the applied force. By comparing coefficients of equations (2.1) and (2.2) it can be seen that for this system

$$\omega_n = \sqrt{\frac{k}{m}} \tag{2.3}$$

$$\xi = \frac{\gamma}{2\sqrt{km}} \tag{2.4}$$

The general solution to equation (2.1) can be either of two types depending on whether the system is being driven by y, or whether the energy it has at $t = 0$ is being allowed to dissipate.

2.2.1   The non-driven oscillator

Three types of solution to equation (2.1) exist depending on whether the system is,

underdamped, $\xi < 1$ critically damped, $\xi = 1$ or overdamped, $\xi > 1$.

The solution for the underdamped case, with $x = x_0$ at $t = 0$, and $x = 0$ at $t = \infty$ is

$$x(t) = x_0 \exp(-\xi\omega_n t)\left[\cos\omega t + \frac{\xi}{(1 - \xi^2)^{1/2}}\sin\omega t\right] \tag{2.5}$$

where $\omega = \omega_n(1 - \xi^2)^{1/2}$.

For resonator sensors, designs are such that $\xi \ll 1$ allowing the approximation

$$x(t) \cong x_0 \exp(-\xi\omega_n t)\cos\omega_n t \tag{2.6}$$

$$x(t) \cong x_0 \exp(-t/2\tau)\cos\omega_n t \tag{2.7}$$

where $\tau = 1/(2\xi\omega_n)$ and is sometimes referred to as the *relaxation* time. This is not to be confused with the relaxation oscillator where the relaxation time of the dissipation of the stored energy determines the oscillation frequency. The form of equation (2.7) is shown in figure 2.1.

2.2.2   The driven oscillator

It is important to realize that when the oscillator is driven sinusoidally the only solution allowed for equation (2.1) is that the oscillation frequency is the same as the driving frequency (this is not true in a nonlinear system). This means that, by using feedback, the frequency of the drive circuitry can be made to follow the resonant frequency of the sensor as it changes and so be used as the output signal. The general solution for a maintained oscillator driven by

$$y = y_0 \sin\omega t \tag{2.8}$$

can be expressed in the form

$$x = x_0 \sin(\omega t + \phi) \tag{2.9}$$

where x is the displacement of the point in the structure where y is input. The phase ϕ is the angle by which the displacement leads the force. To determine the full description of equation (2.9), its derivatives together with equation

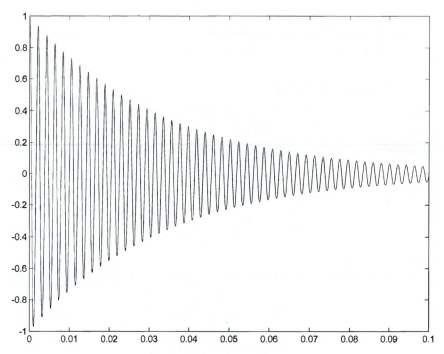

Figure 2.1. Response of a non-driven oscillator.

(2.8) are substituted into equation (2.1):

$$\frac{dx}{dt} = \omega x_0 \cos(\omega t + \phi) \tag{2.10}$$

$$\frac{d^2x}{d^2t} = -\omega^2 x_0 \sin(\omega t + \phi). \tag{2.11}$$

When substituted equation (2.1) becomes

$$\frac{-\omega^2 x_0}{\omega_n^2} \sin(\omega t + \phi) + \frac{2\xi}{\omega_n} \omega x_0 \cos(\omega t + \phi) = y_0 \sin \omega t. \tag{2.12}$$

Gathering terms,

$$\left(1 - \frac{\omega^2}{\omega_n^2}\right) x_0 \sin(\omega t + \phi) + \frac{2\xi}{\omega_n} \omega x_0 \cos(\omega t + \phi) = y_0 \sin \omega t. \tag{2.13}$$

This can be further simplified by using the trigonometric relations

$$\sin(\omega t + \phi) = \sin \omega t \cos \phi + \cos \omega t \sin \phi \tag{2.14}$$

$$\cos(\omega t + \phi) = \cos \omega t \cos \phi - \sin \omega t \sin \phi \tag{2.15}$$

Equation (2.12) then becomes

$$\left[\left(1 - \frac{\omega^2}{\omega_n^2}\right)\cos\phi - \frac{2\xi\omega}{\omega_n}\sin\phi\right]x_0\sin\omega t$$

$$+ \left[\left(1 - \frac{\omega^2}{\omega_n^2}\right)\sin\phi + \frac{2\xi\omega}{\omega_n}\cos\phi\right]x_0\cos\omega t = y_0\sin\omega t. \quad (2.16)$$

Equation (2.16) can only be satisfied if the coefficient of $x_0\cos\omega t$ is zero:

$$\left(1 - \frac{\omega^2}{\omega_n^2}\right)\sin\phi + \frac{2\xi\omega}{\omega_n}\cos\phi = 0. \quad (2.17)$$

This gives the result

$$\tan\phi = \frac{\sin\phi}{\cos\phi} = \frac{-2\xi\omega\omega_n}{(\omega_n^2 - \omega^2)} \quad (2.18)$$

from which can be obtained

$$\cos\phi = \frac{\omega_n^2 - \omega^2}{[(\omega_n^2 - \omega^2)^2 + (2\xi\omega\omega_n)^2]^{1/2}} \quad (2.19)$$

$$\sin\phi = \frac{-2\xi\omega\omega_n}{[(\omega_n^2 - \omega^2)^2 + (2\xi\omega\omega_n)^2]^{1/2}}. \quad (2.20)$$

Equation (2.16) can be re-written

$$\left[\left(1 - \frac{\omega^2}{\omega_n^2}\right)\cos\phi - \frac{2\xi\omega}{\omega_n}\sin\phi\right]x_0\sin\omega t = y_0\sin\omega t \quad (2.21)$$

and substituting equations (2.19) and (2.20) into equation (2.21) gives

$$x_0 = \frac{y_0\omega_n^2}{[(\omega_n^2 - \omega^2)^2 + (2\xi\omega\omega_n)^2]^{1/2}}. \quad (2.22)$$

The full solution for x is therefore

$$x = \frac{y_0\omega_n^2}{[(\omega_n^2 - \omega^2)^2 + (2\xi\omega\omega_n)^2]^{1/2}}\sin\left(\omega t + \tan^{-1}\frac{-2\xi\omega\omega_n}{\omega_n^2 - \omega^2}\right). \quad (2.23)$$

The variation in ϕ with ω is as shown in figure 2.2. The driving force always leads the displacement.

Resonator sensors are usually lightly damped so that the resonant frequency approximates to ω_n. Some form of feedback is used to keep the system operating at its resonant frequency ensuring that the drive frequency ω will always be ω_n. Consequently, the value of ϕ will be maintained at $-\pi/2$. However, in a complex structure, parts will be vibrating with the same frequency but different phases. It may be possible to use these relative phases as an indication of the value of the measurand.

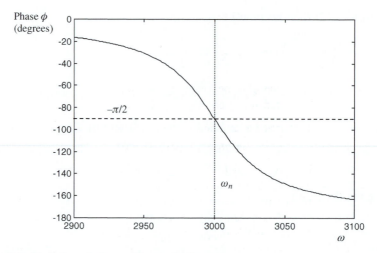

Figure 2.2. Variation of phase with angular frequency.

2.2.3 *Q* factor

A useful parameter when dealing with resonant systems is the Q factor. The definition of Q is

$$Q = \frac{2\pi \times \text{energy stored}}{\text{average energy dissipated per cycle}}.$$

For a mechanical resonator the stored energy will oscillate between potential and kinetic forms. The stored energy will be entirely of the potential form when the velocity is zero and the displacement x is a maximum. From equation (2.6), for the lightly damped, non-driven oscillator, setting the cosine term to a maximum gives

$$x(t) \cong x_0 \exp(-\xi\omega_n t). \tag{2.24}$$

The potential energy E, stored in a linear system of elasticity constant K, is $\frac{1}{2}K(\text{displacement})^2$ giving

$$\text{total stored energy} \cong \tfrac{1}{2}Kx_0^2 \exp(-2\xi\omega_n t). \tag{2.25}$$

The rate of dissipation of energy is $-\mathrm{d}E/\mathrm{d}t$ and differentiating equation (2.25) gives

$$\frac{\mathrm{d}E}{\mathrm{d}t} \cong \xi\omega_n Kx_0^2 \exp(-2\xi\omega_n t). \tag{2.26}$$

Using this with the definition of Q from above gives

$$Q \cong \frac{2\pi \times \tfrac{1}{2}Kx_0^2 \exp(-2\xi\omega_n t)}{\xi\omega_n Kx_0^2 \exp(-2\xi\omega_n t) \times \text{period}} \tag{2.27}$$

$$Q \cong \frac{\pi}{\xi\omega_n \times (2\pi/\omega_n)} \qquad (2.28)$$

$$Q \cong \frac{1}{2\xi}. \qquad (2.29)$$

This useful result enables Q to be easily determined from the system equation via ξ. It can also be seen using equation (2.22) that

$$\frac{\text{the amplitude at resonance}}{\text{the amplitude as the frequency tends to zero}} = \frac{y_0/2\xi}{y_0} \cong Q. \qquad (2.30)$$

This illustrates that, for a high-Q system, the amplitude of the oscillations are Q times larger at resonance than at low frequencies.

It is also instructive to examine the way power is absorbed by the oscillator as the driving frequency is varied. Monitoring the power absorbed by the oscillator when used as a sensor may be useful in monitoring its operational state in order to determine the value of the measurand. The relationship conveniently used to determine the power absorbed, P, is the product of the force and velocity:

$$P = y\frac{dx}{dt}. \qquad (2.31)$$

Substituting equation (2.8) and the differential of equation (2.23) into equation (2.31) gives

$$P = y_0 \sin\omega t \frac{y_0\omega_n^2}{[(\omega_n^2 - \omega^2)^2 + (2\xi\omega\omega_n)^2]^{1/2}} \omega\cos(\omega t + \phi). \qquad (2.32)$$

The cosine term can be expanded using the trigonometric relation

$$\cos(\omega t + \phi) = \cos\omega t\cos\phi - \sin\omega t\sin\phi \qquad (2.33)$$

so that equation (2.32) becomes

$$P = \frac{y_0^2\omega_n^2}{[(\omega_n^2 - \omega^2)^2 + (2\xi\omega\omega_n)^2]^{1/2}} \omega[\sin\omega t\cos\omega t\cos\phi - \sin^2\omega t\sin\phi]. \qquad (2.34)$$

The time average of $\sin\omega t\cos\omega t$ over a period is zero and of $\sin^2\omega t$ is $\frac{1}{2}$. Equation (2.32) then becomes

$$P = \frac{-\frac{1}{2}y_0^2\omega_n^2\omega\sin\phi}{[(\omega_n^2 - \omega^2)^2 + (2\xi\omega\omega_n)^2]^{1/2}}. \qquad (2.35)$$

Substituting for $\sin\phi$ using equation (2.20)

$$P = \frac{y_0^2\omega_n^3\xi\omega^2}{(\omega_n^2 - \omega^2)^2 + (2\xi\omega\omega_n)^2}. \qquad (2.36)$$

The variation in absorbed power, P, with ω is shown in figure 2.3.

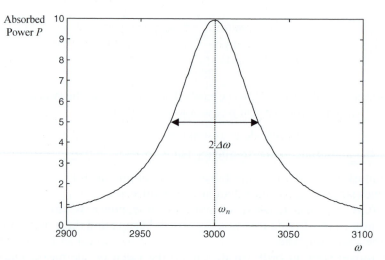

Figure 2.3. Variation in absorbed power with angular frequency in a high-Q resonator.

Engineers often find the frequency spread, from where the peak occurs to the points where the power has dropped by half, a useful concept to describe the narrowness of the resonance peak. Using equation (2.36) to determine the half-power points the denominator $[(\omega_n^2 - \omega^2)^2 + (2\xi\omega\omega_n)^2]$ is doubled.

To evaluate this, first consider the power at the peak where $\omega = \omega_n$. This reduces the denominator $[(\omega_n^2 - \omega^2)^2 + (2\xi\omega\omega_n)^2]$ to $(2\xi\omega_n^2)^2$. Thus, for half power, the denominator must be equal to twice $(2\xi\omega_n^2)^2$:

$$[(\omega_n^2 - \omega^2)^2 + (2\xi\omega\omega_n)^2] = 2(2\xi\omega_n^2)^2. \tag{2.37}$$

Re-arranging,

$$(\omega_n^2 - \omega^2)^2 = 2(2\xi\omega_n^2)^2 - (2\xi\omega\omega_n)^2 \tag{2.38}$$

which can be re-written as

$$(\omega_n - \omega)^2(\omega_n + \omega)^2 = 2(2\xi\omega_n^2)^2 - (2\xi\omega\omega_n)^2. \tag{2.39}$$

Putting $\Delta\omega = \omega_n - \omega$ and making the assumption that $\omega_n \cong \omega$ gives

$$\Delta\omega 2\omega_n = 2\xi\omega_n^2 \tag{2.40}$$

$$\Delta\omega = \xi\omega_n \tag{2.41}$$

and, using equation (2.29),

$$\Delta\omega = \frac{\omega_n}{2Q}. \tag{2.42}$$

This provides a further description of Q,

$$Q = \frac{\omega_n}{2\Delta\omega} \tag{2.43}$$

where $2\Delta\omega$ is the full bandwidth of the frequency spread at the half-power points. This indicates that a high-Q system has a narrow resonant peak.

To conclude this section, the advantages of high Q are summarized as:

- low dissipation of energy per cycle so that little power is needed to maintain oscillation;
- a narrow frequency band over which resonance occurs giving the sensor high resolution;
- greater immunity to loading effects from sense and drive circuitry.

It should be noted that some oscillatory sensors operate over a small range of frequencies around resonance. For example, the sensor may sweep through a range of frequencies in order to collect information about the measurand. In such systems low Q is preferable so that constant amplitude exists over the working frequency range.

Sections 2.3 and following illustrate some resonant sensor designs and their applications.

2.2.4   Driving techniques

As described above the measurand modifies the natural frequency of the resonator sensor and this can be output as a representation of the measured value. To maintain the oscillation at this variable resonant frequency a feedback loop is required. A detector unit must measure some parameter that indicates the status of the sensor. A control unit then decides whether the sensor is at resonance and the appropriate output is then fed to a power oscillator. The power oscillator supplies the energy required to drive the sensor and also provides the sensor output.

There are three ways by which feedback may be employed.

(a) Monitoring of the amplitude of the resonator output and adjusting the drive frequency so as to maintain this at a maximum.
(b) Monitoring the phase difference between the driving signal and the displacement of the sensor. The drive signal leads the displacement by $\pi/2$ at resonance.

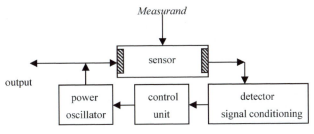

Figure 2.4. Block diagram of a resonator sensor drive system.

(c) Monitoring the power absorbed by the resonator, which is a maximum at resonance.

Readily available integrated circuits can be used to construct the system. Typically, these might include comparators, voltage controlled oscillators, phase-locked loops and gating circuits.

Matlab® programs

The following M-file was used to draw figure 2.1.

```
clear, clf
wn=3000; dampratio=0.01; T=1/(2*wn*dampratio);
c=(1-dampratio^2)^0.5;
w=wn*c;
time= [0:0.0001:0.1];

for   j=1:length(time);
      t=time(j);
      A=exp(-dampratio*wn*t);
      cosine=cos(w*t);
      sine=sin(w*t);
      B=dampratio/c;
      G(j,:)=A*(cosine+B*sine);
end
figure(1), plot(time,G,'k')
```

The following M-file was used to draw figure 2.2 (labelled figure (1) in the file) and figure 2.3 (labelled figure (5) in the file).

```
clear, clf
wn=3000; dampratio=0.01; T=1/(2*wn*dampratio);
FREQ= [2900:1:3100];
for   j=1:length(FREQ);
      w=FREQ(j);
   A=(wn^2-w^2);
   B=w/T;
   D1=A.^2;
   D2=B^2;
   D3=(D1+D2);
   D=D3^0.5;
   cosine=A./D;
   sine=-B./D;
   tangent=-B./A;
   G(j,:)=acos(cosine);
   H(j,:)=asin(sine);
```

```
K(j,:)=atan(tangent);
M(j,:)=cosine;
power(j,:)=-0.199*wn^2*sine./D;
```
end
```
figure(1), plot(FREQ,-G*180/pi)
figure(2), plot(FREQ,H*180/pi)
figure(3), plot(FREQ,K*180/pi)
figure(4), plot(FREQ,M)
figure(5), plot(FREQ,power)
```

2.3   Vibrating wire sensors

This device is the simplest of all the resonator sensors. Assuming that the rigidity of the wire to bending is negligible, the frequency of transverse oscillations is

$$\omega_n = \frac{\pi}{l}\sqrt{\frac{T}{\mu}} \qquad (2.44)$$

where T is the tension, μ is the mass per unit length and l is the length of wire.

Sensors have been developed using wires made from the metals tungsten, indium and high tensile steel as well as alloys such as Elinvar.

Excitation is achieved by using an a.c. current through the wire, which is mounted in a permanent magnetic field. The terminal impedance varies strongly through resonance so the drive circuitry is designed to produce the required frequency by tuning to this condition.

The areas of application are in measuring force, mass, strain and pressure. With good design practice and protection of the wire from contamination, zero point stability, and hysteresis can be as low as $\pm0.02\%$. Expansion of the wire due to temperature produces errors of order $10\,\mathrm{ppm}/^\circ\mathrm{C}$. Marshall and Hunter (1980) reported using the vibrating wire to measure pressure. A commercial device was available from Foxboro. Figure 2.5 shows a double diaphragm arrangement.

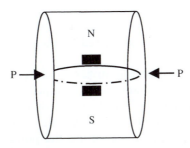

Figure 2.5. Differential pressure cell.

2.4   Torsionally vibrating rod

A liquid density sensor (figure 2.6) was developed by Langdon (1980) at GEC for use in lead–acid batteries. The head unit contains a piezoelectric driver to vibrate the rod, which is made of glass. Power comes from the battery itself, making the unit self-contained. Density measurements are claimed to be as accurate as float hydrometers, having an error of ±0.2%.

For a rotational second order system equation (2.2) can be re-written as

$$I\frac{d^2\theta}{dt^2} + \chi\frac{d\theta}{dt} + c\theta = \tau \tag{2.45}$$

where I is the moment of inertia of the rotating component, χ is the rotational damping constant, c is the spring restoring torque constant and τ is the applied torque. By comparing terms with equation (2.3) the undamped natural frequency can be seen to be

$$\omega_n = \sqrt{\frac{c}{I}}. \tag{2.46}$$

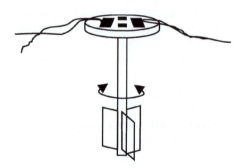

Figure 2.6. Torsion rod liquid density sensor.

2.5   Longitudinally vibrating plate

An established, non-resonating, method for on-line viscosity measurement uses a vibrating sphere driven at a fixed frequency of 200 Hz (Matusik and Scarna 1981).

A more compact sensor for on-line viscosity measurement has been devised using a thin rectangular plate clamped at its centre (Borovikov 1976) (figure 2.7). It may be driven and monitored by magnetostrictive devices to produce longitudinal resonance. The damping, measured either by switching off the power or measuring the power loss, is then used to determine the viscosity. Alternatively, the phase shift ϕ between driving

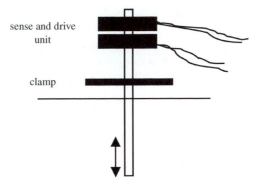

Figure 2.7. Viscosity measurement using a longitudinally resonating plate.

force and displacement can be used where

$$\phi = \tan^{-1} \frac{2\omega_n \alpha}{\omega^2 - \omega_n^2} \tag{2.47}$$

with α approximately proportional to the square root of the dynamic viscosity. The most significant disadvantage of the sensor is that penetration of the wave motion into the liquid is very poor. At the typical 20 kHz drive frequencies penetration is $<10\ \mu m$, so probe cleanliness is essential. The non-resonating device operating at the much lower frequency does not suffer from poor penetration.

2.6 Flexural resonators

A number of designs have been put forward for flow measurement using a flexing structure.

Flexing vane

MacDonald (1983) produced a device (figure 2.8) having a small vane (metal or ceramic can be used) placed longitudinally in the pipe. It was driven into flexural resonance at about 1 kHz by a centre-mounted transducer. Phase

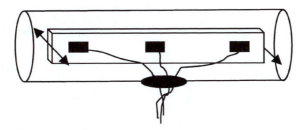

Figure 2.8. Flexing vane mass flow rate sensor.

differences between pickups at the two ends of the vane were used to indicate mass flow rate. Prototypes were built by GEC using a vane size of 150 mm × 20 mm which gave linear mass flow rate measurement for velocities of 0–4 ± 0.02 m s^{-1}. As is common in such systems, protection of the sensing elements from contaminants that adhere is important and entrained gas can also give false readings.

Flexing pipe

In this design the whole pipe section is set into resonance. The resonance process can be likened to the vibrating wire sensor, the expression for the undamped natural frequency ω_n being of the form of equation (2.3). The resonant frequency is determined by the total mass in the device (including the pipe itself). This makes the sensor useful in multi-phase or multi-component applications. The Sarosota company have developed an on-line density meter (figure 2.9(a)), in which the expression for the density is given as

$$\rho = \frac{A}{\omega_n^2} + \frac{B}{\omega_n} + C. \tag{2.48}$$

A different approach is to use the Coriolis effect. Coriolis showed that a moving object in an accelerating reference frame experiences a side thrust. The accelerating reference frame in this application is the oscillating pipe. The vector equation relating the Coriolis force is

$$\boldsymbol{F_C} = 2m\omega\Lambda\left(\frac{\mathrm{d}\boldsymbol{r}}{\mathrm{d}t}\right) \tag{2.49}$$

where m is the mass of the object moving at an angular velocity ω whilst changing its radius vector $\boldsymbol{r}$. $\boldsymbol{F_C}$ and $\boldsymbol{r}$ are perpendicular. The device measures mass flow. Because large lengths of pipe are needed the pipe is often bent into a U-tube (figure 2.9(b)). Schlumberger and Micro Motion are among a number of companies now marketing straight and U-shaped systems.

The Coriolis force causes a tiny twist to the pipe which can be detected by a pair of position sensors or strain gauges.

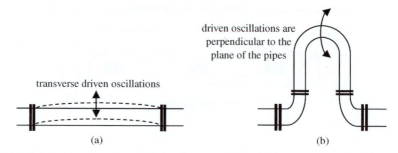

driven oscillations are perpendicular to the plane of the pipes

transverse driven oscillations

(a)   (b)

Figure 2.9. Examples of vibrating pipe flow rate sensors.

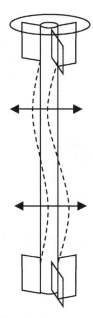

Figure 2.10. Flexing tube liquid level sensor.

Flexing tube

A liquid level sensor has been devised using a flexing tube (figure 2.10) 300 mm long and 6 mm in diameter (Langdon 1983). The effective flexural mass of the tube is altered by the liquid surrounding it. This causes the resonant frequency of about 5 kHz to decrease as the depth of liquid increases. The response is linear apart from when the liquid passes through a tube nodal point. This can be overcome by either

1. driving the device sequentially at different resonant frequencies, or
2. by 'plane polarizing' the vibration pattern using vanes on the tube and driving it at two different resonant frequencies at the same time.

In either method the nodes appear at different points. By combining the output responses the nonlinearities can be removed.

These devices are mass level sensors and as such the density of the liquid affects the output and may need to be measured separately. However, in some applications, such as aircraft re-fuelling, this is the desired quantity.

2.7 Vibrating cylinder

A thin walled metal cylinder mounted axially within a pipe can be used as an on-line densitometer (figure 2.11) (Agar 1967, 1976). It is held rigidly at one

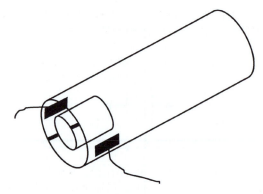

Figure 2.11. Vibrating cylinder on-line densitometer.

end and can be excited electromagnetically by external coils if made of ferromagnetic material. If the liquid is corrosive a ceramic tube is used instead with piezoelectric sense/drive devices bonded to it. Being a thin tube, it does not restrict the flow to any extent. Liquid passes over both surfaces so radial pressure differences are minimal. Sensor dimensions should be much smaller than the wavelength so that acoustic coupling is low. Under this restriction the fluid behaves as incompressible and ω_n is unaffected.

A further application of the vibrating cylinder is in the construction of a rate gyro (figure 2.12). G H Bryan in 1890 observed that the vibration pattern in a vibrating wineglass lagged the angle of rotation. This is a further example of the Coriolis effect. The development of a sensor using this phenomenon has been reported by Langdon (1982). The cylinder, which can be of metal or ceramic, is driven at its fundamental flexural mode by piezoelectric transducers producing four nodes in symmetrical positions around the cylinder. When rotated these nodes move and their angular change, monitored by the piezoelectric transducers, is used to give an output proportional to rotation rate.

Figure 2.12. Vibrating cylinder rate gyro.

A commercial device has been developed by GEC Marconi Avionics with a sensitivity of $0.01° \, s^{-1}$ operating up to $480° \, s^{-1}$. An automotive version has been developed from this by Bosch (see Reppich and Willig 1995). Johnson (1995) has reported a similar device developed by British Aerospace for use in the automotive industry and is in commercial production.

2.8 The tuning fork

An invasive liquid density sensor is available from Schlumberger using tuning fork-type tines. It will operate in static tanks or in pipelines where flow rates are below $3 \, m \, s^{-1}$. The tines are driven and sensed by piezoresistive devices and the resonant condition is maintained by monitoring the phase. The relationship between the density ρ and the natural frequency is given by

$$\rho = K_0 + K_1 \tau + K_2 \tau^2 \tag{2.50}$$

where K_0, K_1 and K_2 are constants and τ is the time period of the resonant frequency.

Ichinose and Terada (1995) have reported the development of an angular rate sensor for automotive application. A miniature tuning fork structure was designed using finite element techniques. The resonant frequency is approximated as

$$\omega_n = \frac{\alpha^2 t}{l^2} \sqrt{\frac{\rho}{E}} \tag{2.51}$$

where α is the eigenvalue, t is the thickness, l is the length, E is Young's modulus and ρ is the density. The effects of the Coriolis force on the resonating tines were measured using piezoelectric sensors. Signal recovery was enhanced by synchronous detection.

2.9 The double tuning fork structure

Structures of this type have been developed for on-line density measurement (figure 2.13). The flow pipe bifurcates to form the double tuning fork configuration. The contra motion of the side-sections tends to cancel any bending forces that would otherwise be presented at the mountings. However, axial stresses at the mountings can affect the resonant frequency so flexible couplings have to be used. The resonant frequency is related to the density ρ by

$$f = f_0 \sqrt{\frac{\rho_0}{\rho_0 + \rho}} \tag{2.52}$$

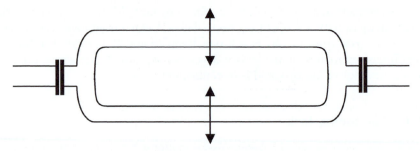

Figure 2.13. On-line density measurement using a bifurcated pipe.

where ρ_0 is a constant. Errors of 0.1% or better can be achieved (Dress 1983).

A commercial pressure sensor using the double tuning fork structure is now available from Yokogawa of Japan. The vibrating element is of solid Elinvar alloy. In this configuration the modulus of elasticity E changes proportionally with axial force. The force comes from a diaphragm at one end of the vibrator. The resonant frequency–pressure relationship is non-linear so a microprocessor is used with a look-up table to output a linear response. The value of Q is $>10^4$ and errors are typically $\pm0.02\%$ of the reading. With suitable temperature compensation (most error is said to be due to thermal expansion of the diaphragm) zero and sensitivity drift can be as low as $\pm0.001\%$ per °C of full scale.

Interest in tuning fork-type structures still continues. Randall *et al* (1997) have reported a pressure sensor using a metallic, triple-beam structure. It is stated that loss of energy due to clamping is minimized in the symmetrical structure used. Beeby *et al* (2000) describe the fabrication and testing of a double tuning-fork, resonator microsensor constructed in silicon.

2.10 Bulk acoustic wave resonator sensors

Since the early 1970s quartz crystal systems have been commercially available for monitoring thin film deposition in vacuum evaporators. The exposed quartz is part of an oscillating circuit and usually operates in thick-ness-shear-mode. The resonant frequency is modified by the gain in mass. Such devices are sometimes referred to as microbalances. Unfortunately, temperature also affects the resonant frequency. Indeed, the technique can be used for temperature measurement and commercial thermometers are available with sensitivities of 0.001 °C. By combining both sensors to produce a temperature compensated thin-film monitor it is possible to monitor deposition of material to atomic layer resolution (figure 2.14).

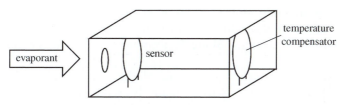

Figure 2.14. Thin film thickness-monitor using quartz resonators.

A typical commercial AT-cut quartz oscillator is a disk about 10–16 mm in diameter or a square or rectangle of similar size. The Sauerbrey equation relates the frequency shift to the change in mass,

$$\Delta f = \frac{-f_0^2 \Delta m}{\rho k A} \tag{2.53}$$

where ρ is the density, k is the frequency constant and A is the surface area of the quartz crystal. Substituting these constants for quartz gives

$$\Delta f = \frac{-2.25 \times 10^{-2} f_0^2 \Delta m}{A} \tag{2.54}$$

with A expressed in m^2 and m in kg. Operating frequencies are usually 5, 9 or 10 MHz. At 10 MHz a sensitivity of 0.3 Hz per nanogram is achievable.

Various surface coatings have been developed for adsorbing a specific gas or vapour. In this way the microbalance becomes a qualitative and quantitative sensor for a particular chemical species. Finding highly selective coatings is not easy.

Biologically active coatings have also been produced to create specific biosensors for use in the gas/vapour phase as well as in solution. The attractive feature with biological reagents is that they are extremely specific. However, the construction methodology for biological systems is much more difficult and often the sensor can only be used once.

2.11 Thick films

A number of the sensors described above have piezoelectric drive and sense transducers. Commonly, fabrication of these components is by some form of thick film process. Films can be classified by order of magnitude of their thickness: ultra-thin <0.1 µm, thin 0.1–1.0 µm, thick 1.0–100.0 µm.

Thin film production often uses techniques developed by the integrated circuit industry. As described later in chapter 3, this can be ideal when sensor and electronics are combined on a single chip. However, some coating materials cannot withstand the harsh production processes used. Also, manufacture in silicon foundries is expensive and may not be viable for small-scale production. In these cases other techniques are more appropriate for both thin and thick film production. These include electrochemical deposition, the Langmuir–Blodgett technique, spin casting and screen printing.

Of these, screen printing was one of the earliest techniques adapted for thick film production. Sensors made by screen printing have been in development for some time (see Brignell *et al* 1987). The piezoresistance strain gauge has enjoyed commercial success, first marketed about two decades ago.

Screen printing is an adaptation of silkscreen printing used in fine art for pattern layering in textile and paper design and in some paintings. Here, a synthetic fabric (silk is obsolete now) is stretched on to a rigid frame. A pattern is laid down on the fabric by the artist and hot wax poured on to make parts of the fabric impermeable to the printing ink. Additionally, once the wax has set patterns can be carved forming permeable pathways for the ink to reach the workpiece. The screen is then held firmly against the workpiece using a press. The ink is spread uniformly over the fabric by a roller and so penetrates through to the workpiece forming the desired design. The process can be repeated with different designs and ink colours to produce a composite.

For sensor application the desired layout for the active material is formed on the screen fabric using photolithography so as to create a mask. This might be, for example, a bridge circuit configuration. The 'ink' is squeezed through the mask to form a film having a thickness typically in the range 25–100 μm. The film is then dried by low temperature heat treatment. Following this stage high temperatures are used to cause sintering of the ingredients of the paste and wetting of the substrate so as to form good adhesion. The firing profile is critical to the production quality.

The screen printing technique is not as highly specified as that of the semiconductor foundry systems and consequently these sensors tend to have poorer manufacturing tolerances. Researchers and manufacturers produce recipes, sometimes of secret formulation, for these printable pastes. The general formulation is for the active material to be suspended in an organic solvent together with some form of matrix material (see Baudry 1987 and Morton *et al* 1991). Here a piezoelectric film based on lead zirconate titanate (PZT) was used. Some materials which are not naturally piezoelectric can be polarized to make them so. This usually involves gently heating the film and allowing it to cool in the presence of a strong electric field.

Interest in thick film sensors is still high, principally because of the lower costs in making small, hybrid devices and because operating temperatures can be higher than for comparable thin film sensors. As with other sensor types described in chapter 3, advances are dependent on organic and physical chemistry for the development of film materials.

References

Agar J 1967, 1976 *Apparatus for Measuring the Density of a Fluid* UK Patents 1175586, 1542564

Baudry H 1987 'Screen printing piezoelectric devices' *Proc. 6th European Microelectronics Conference* (*Bournemouth*) pp 456–463

Beeby S P, Ensell G and White N M 2000 'Microengineered silicon double-ended tuning-fork resonators' *Engineering Science and Education J.* December

Borovikov A P 1976 'Measurement of viscosity of media by means of shear vibration of plane resonators' *Inst. Exp. Tech.* **19**(2) 223–224

Brignell J E, White N M and Cranny A W J 1987 'Sensor applications of thick film technology' *Proc. IEE I* **135**(4) 77–84

Dress W B 1983 'A high resolution ultrasonic densitometer' *IEEE Ultrasonics Symposium* pp 287–290

Ichinose T and Terada J 1995 'Angular rate sensor for automotive application' *Sensors and Actuators* **SP-1066** pp 49–56

Johnson B 1995 'Vibration rotation sensors' *Sensors and Actuators* **SP-1066** pp 41–47

Langdon R M, 1980 'Liquid level measurement' UK Patent GB 2067756A

Langdon R M 1982 *Marconi Review* Fourth Quarter pp 231–249

Langdon R M 1983 'Vibrating transducers in measurement' *Chem. Engineer* **397** 33–35

MacDonald 1983 'A vibrating vane mass flow meter' *1st Euro. Conf. Sensors and their Applications, UMIST* pp 58–59

Marshall J K and Hunter P 1980 *Strain* (*GB*) **1** 37–40

Matusik F J and Scarna P C 1981 'Latest instrument makes on-line viscosity control of slurries possible' *Control Engineering* **28** 116–118

Morton B, De Cicco G, Gandolfi A and Tonelli C 1991 'Advances in ferrelectric thick film materials and sensors' *Proc. 8th European Microelectronics Conference* (*Rotterdam*) pp 392–399

Randall D S, Rudkin M J, Cheshmehdoost A and Jones B E 1997 'A pressure transducer using a metallic triple-beam tuning fork' *Sensors and Actuators A: Physical* **60**(1–3) 160–162

Reppich A and Willig R 1995 'Yaw rate sensor for vehicle dynamics control system' *Sensors and Actuators* **SP-1066** pp 67–76

Sarosota Automation Instrumentation Division 1985 Technical Information on FD 800 series density meters

Exercises

1. (i) State two advantages and two disadvantages to resonator sensors.
    (ii) Distinguish between relaxation and harmonic oscillators.
    (iii) Give three advantages to having high Q in resonant systems.

2. (i) Describe three approaches by which the resonant frequency might be tracked.
    (ii) Using functional blocks, design systems for these three different approaches to tracking.

3. The general equation for a damped second order system is

$$\frac{1}{\omega_n^2}\frac{d^2x}{dt^2} + \frac{2\xi}{\omega_n}\frac{dx}{dt} + x = y.$$

Show, by using the Laplace transform or similar technique, that the solution for a lightly damped, non-driven system with zero starting conditions is

$$x(t) = x_0\exp(-\xi\omega_n t)\left[\cos\omega t + \frac{\xi}{(1-\xi^2)^{1/2}}\sin\omega t\right].$$

4. For a certain resonator sensor when the power is turned off the displacement x declines as

$$x(t) = x_0\exp(-\xi\omega_n t)\cos\omega_n t.$$

Given that the stored potential energy is obtained using $\frac{1}{2}Kx^2$ determine the maximum potential energy as a function of t and hence show that $Q \cong 1/(2\xi)$.

5. A certain resonator sensor for measuring liquid density on-line can be modelled by the equation

$$1.12\times10^{-7}\frac{d^2x}{dt^2} + 7.4\times10^{-7}\frac{dx}{dt} + x = y.$$

(i) Determine the undamped natural frequency ω_n, the damping ratio ξ and the quality factor Q.
(ii) If the density of the liquid is related to the resonant frequency by

$$\rho = \frac{\alpha}{\omega_n^2} + \frac{\beta}{\omega_n} + \gamma$$

where α, β and γ are all constants, derive an expression for the sensitivity.

6. A certain pressure sensor is constructed using a vibrating wire. Given that the frequency f of the wire goes as $f = 1/(2l)\sqrt{T/\mu}$, determine the frequency shift if the temperature decreases by 15 °C. The original frequency is 50 kHz and the coefficient of linear expansion for the wire material is $1.4\times10^{-5}\,K^{-1}$. Assume that the change in the wire tension is linearly related to the change in length and that the actual length of the wire remains fixed.

7. (i) With the aid of a diagram explain the Coriolis effect, indicating the relative orientations of the parameters involved.
(ii) Coriolis liquid flowmeters can be constructed using either straight or U-shaped pipes. Draw a diagram for each type indicating the force direction. How might this force be measured?

Chapter 3

Semiconductor-based sensors

3.1 Introduction

The development of the silicon-based technologies in the latter half of the twentieth century has undoubtedly had a major impact on societies throughout the world. Components fabricated from silicon can be found in every type of electronic device. Such devices have themselves enabled the sharp increase in the rate of change of technological development and its implementation. Although new semiconductor materials are coming into more widespread use the vast majority of the devices produced are still silicon based. The maturity of silicon foundry technology provides relatively cheap routes to manufacturing the now vast range of products from the simplest of discrete components to complex VLSI structures. As well as the proliferation of components for digital applications, the use of power electronic devices increases steadily and the explosive growth in the use of cellular phones has caused rapid development of microwave-frequency components.

The development of digital component technology is no longer the forefront of research. The new challenge is the development of microsensors to interface the computer directly to the real world. The use of microprocessors with sensors is already well established in the so-called *smart* sensors (see section 5.5). This trend is set to continue with the combining of sensor and circuitry on a single chip. For many applications the concept of a sensor and its associated signal conditioning/processing circuitry being made together in a single miniature package is very attractive.

Although single-crystal silicon structures can be used to sense light, strain, temperature, magnetic and electric fields the sensitivities may not be as high as that of more traditional sensors. An example of this is the application of the Hall effect for the measurement of magnetic field. The discrete sensor can be constructed from a small plate of a III–V compound (formed from elements taken from groups III and V in the periodic table) such as indium arsenide, indium antimonide or gallium arsenide, whereas a magnetometer fabricated in silicon using bipolar technology is less sensitive. However, a recent development is to use NMOS technology to produce an ultra-thin slab of around 10 nm so that a larger potential difference is generated. Even though carrier mobility is lower than in a bipolar device, sensitivity is increased by a factor of 20.

The design problems for fabricating semiconductor sensors can be significant. For example, provision may be needed for an input filter, crucial in keeping all but the desired measurand away from the exotic sensing element. Harsh environments of temperature, vibration or pressure, or the presence of substances that would poison the sensor, can also pose difficulties for design engineers.

Interdisciplinary teams are needed for the design, development and fabrication processes. If the sensor can be fabricated entirely in silicon it can be manufactured at the same time as all the other components on the chip using the same masks and stages. However, if, rather than adopting available silicon foundry plant, non-standard manufacturing techniques are required to produce a hybrid, the costs may mean that a proposed device will only be economically viable if high volumes can be sold.

3.2   Mechanical microsensors

A number of new materials are finding applications in mechanical microsensors. Silicon has been used for several decades and is attractive because the manufacturing technology is well developed to produce large, extremely pure, single crystals. These can be doped, etched or machined as desired. The mechanical properties of silicon are also very attractive to sensor design engineers. It has

- high tensile strength
- good fatigue strength
- high Young's modulus, similar to steel
- good strength/weight
- good hardness
- good resistance to deforming plastically so creep and hysteresis effects are minimal.

Construction is usually in single-crystal silicon. However, polycrystalline silicon is more flexible and so is sometimes preferred for deflecting structures which are in turn constructed on single-crystal substrates. A disadvantage is that polycrystalline silicon is more prone to creep and hysteresis.

Sensor developments using *bulk micromachining* began in the early 1960s. A single crystal wafer of silicon is selectively processed throughout its bulk. This may involve architectural dimensions of less then 1 μm or up to the full thickness of the wafer of 500 μm. Most commercial microsensors currently are of the bulk-machined type.

Since the early 1980s *surface micromachining* has attracted a lot of attention. As the name suggests, the sensing element is a thin layer within the surface. The advantages are a reduction in sensor size of an order of magnitude compared with a bulk micromachined device and the relative ease of adding integrated circuitry adjacent to the sensing element. The main disadvantage is that, because of their smaller size the sensitivities may not be as high as can be achieved using bulk micromachining. Consequently, additional circuitry has to be added to the substrate to amplify the response.

Mechanical microsensors for the measurement of strain, pressure, acceleration and gas flow are commercially available. The general method of operation is that the measurand affects some mechanical feature of the sensor and a secondary sensor using capacitance, piezoresistance or, to a lesser extent piezoelectricity, provides the electrical signal.

3.2.1   Using silicon to measure strain

p–n junction devices

The physics of the p–n junction is known to be sensitive to stress. The bandgap energy changes proportionally with stress. It has been suggested that diodes, bipolar transistors and MOSFETs might provide miniature stress sensors. However, stresses need to be high and, as yet, reproducibility has been difficult to achieve. No commercial application is available at the time of writing.

Piezoresistance devices

Unlike metallic strain gauges, where the change of the bulk shape causes change in the resistance, in semiconductor gauges it is the change of the resistivity by the strain that is the dominant term. The gauge factor G of a strain gauge is given by

$$G = 1 + 2\nu + \frac{\Delta\rho}{\rho e_\mathrm{L}} \qquad (3.1)$$

where ν is Poisson's ratio, ρ is the resistivity, and e_L is the longitudinal strain.

The fractional change in resistivity per strain, $\Delta\rho/\rho e_\mathrm{L}$, is the *piezoresistivity*. In metals the piezoresistivity is approximately 0.3 and, together with a typical value for ν of 0.5, gives G a value of about 2. However, in silicon based semiconductors the piezoresistivity can be >100. This becomes the dominating term in the gauge factor equation. Semiconductor strain gauges are available with G values of up to 150. Some other advantages of semiconductor strain gauges are as follows.

- The sensing element(s) can be tailored to the application easily in order to produce maximum sensitivity.
- Matching sets of sensors required for bridge circuitry techniques are easily produced.
- Miniaturization is easily achieved.
- Mass production is straightforward using the silicon industry techniques.
- n- and p-type doped devices are available having opposite sign piezoresistive coefficients. It should also be noted that different values of coefficients exist for the different crystal orientations.

The main disadvantage to semiconductor, piezoresistive devices is their significant temperature sensitivity. The piezoresistive coefficient is temperature sensitive. This is reduced if doping is increased but this also reduces the piezoresistive coefficient. It must also be remembered that the semiconductor material itself has a high temperature coefficient of resistance, making the zero-setting resistance fluctuate. These two effects can be minimized by using a compensation technique such as a bridge network. The zero value of a device is also subject to temperature-dependent drift due to mechanical stresses incorporated during manufacture. Careful design is required to minimize these.

3.2.2   Pressure gauges

The growth rate in the use of miniature pressure sensors has been phenomenal in recent years. The automotive industry alone has seen an increase of 20% per annum in the past decade. The earliest application was probably in 1989 for monitoring exhaust gas re-circulation. More recently devices have been incorporated for measuring manifold, tyre, fuel and hydraulic fluid pressures. Gauge types are classified according to the pressure reference:

- *absolute*: compared with vacuum;
- *differential*: pressure is measured between two points in the system;
- *gauge*: compared with atmospheric pressure.

The piezoresistive phenomenon described above is commonly used as the basis of the sensing process. Bulk-machining is used on a single-crystal silicon plate or circular diaphragm of order $10\,\mu\text{m}$ thick. Piezoresistors are constructed within the diaphragm inner surface which is of order $1\,\text{mm}^2$. Their layout is governed by the strain geometry of the diaphragm as it deforms under the applied pressure. For a circular diaphragm two strain directions need to be considered, one along the radial direction and the other normal to the radius resulting from *hoop* stress (figure 3.1). Both stress relationships are a function of radius. Dalley and Riley (1978) present the strain equations for the diaphragm:

$$\text{radial stress } \varepsilon_{\text{r}} = \frac{3P(1-\nu^2)}{8Et^2}(R_0^2 - 3r^2) \tag{3.2}$$

Figure 3.1. Strain orientations in a simple diaphragm.

$$\text{tangential stress } \varepsilon_t = \frac{3P(1-\nu^2)}{8Et^2}(R_0^2 - r^2) \qquad (3.3)$$

where E is the Young's modulus, ν is Poisson's ratio, t is the thickness of the diaphragm material, R_0 is the diaphragm radius, r is some point along the radius, and P is the differential pressure.

It can be seen that ε_r changes sign at the point where $R_0^2 = 3r^2$, whereas ε_t is always positive. Thus by careful positioning of four strain gauges, two in compression and two in tension, high sensitivity and temperature compensation can be achieved.

Piezoresistive pressure gauges have been available commercially since the early 1980s. Typical parameter values are

range	0–100 kPa
sensitivity	3 mV/kPa with a 10 V power supply
frequency response	100 kHz
nonlinearity and hysteresis	±0.5%

They cannot be used in high temperature or pressure systems directly and, with certain noxious materials, without protection.

Kress *et al* (1995) describe an integrated silicon pressure sensor for automotive applications such as manifold intake pressure, fuel tank and brake-line hydraulic pressure. Not only does the device contain on-board signal conditioning but also an electronic 'trimming' system. This trimming process corrects for production tolerances including sensitivity and is performed by sending a digital compensation code word to the chip. The applied voltage is such as to irreversibly store the sensor characteristic. Previous designs have used the more expensive laser trimming technique. The pressure range is 20–115 kPa (absolute) and the output is 0.4–4.65 V using a 5 V supply. An isolation technique is used enabling the device to operate over the full automotive temperature range of −40 °C to 125 °C.

A pressure sensor having higher sensitivity has been reported using resonance, Greenwood (1988). A shuttle structure was machined out of single-crystal silicon. The supporting struts for the shuttle were connected to the diaphragm. As the diaphragm deformed the changes in stress on the shuttle supports changed the mechanical resonant frequency. High long-term stability was achieved. Recent work by others on similar devices has shown that high precision is possible, making it attractive for commercialization.

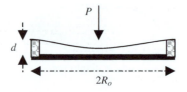

Figure 3.2. Conventional capacitance pressure gauge.

However, it is likely to be marketed as a high-cost device due to the complexity of fabrication.

Pressure gauges are also manufactured using capacitance as the secondary sensor. This approach has been used in non-microsensors since the emergence of the instrumentation industry. In these traditional devices the whole of the diaphragm acts as one of the plates of a capacitor. Shown in figure 3.2 is this type of capacitive pressure sensor. One plate is a fixed metal base, the other is a flexible circular diaphragm which is clamped around its circumference. The diaphragm is deformed as shown by the applied pressure P.

The capacitance at zero pressure can be determined using the formula for a parallel plate capacitor,

$$C = \frac{\varepsilon_0 \pi R_0^2}{d}. \tag{3.4}$$

The deformation of the diaphragm means that the average separation of the plates is reduced and the resulting increase in capacitance ΔC is given by Neubert (1975) as

$$\frac{\Delta C}{C} = \frac{(1 - \nu^2) R_0^4}{16E \, \mathrm{d}t^3} P \tag{3.5}$$

where, E is the Young's modulus, ν is Poisson's ratio and t is the thickness of the diaphragm material.

The microsensor versions of this device use a silicon diaphragm and evaporated metal film to form the upper and lower capacitor plates. Some versions do not use the whole diaphragm as one of the capacitor plates. Instead sets of miniature capacitors are fabricated within the gap, those at the edges being unaffected by the pressure providing a reference to give temperature compensation.

3.2.3 Accelerometers

The growth in the commercial application of the miniature accelerometer has been almost as great as that of the miniature pressure sensor. There has always been a demand for accelerometers but in recent years their usage has soared. Again the automotive industry has shown rapid take-up for

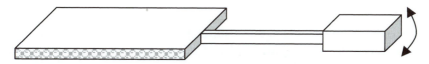

Figure 3.3. A simple cantilever accelerometer.

installation in air-bag, active suspension and breaking systems (see Gardener *et al* 2001).

An accelerometer is a device to measure acceleration or, in some configurations, vibration (figure 3.3). It consists of an inertial mass m, which undergoes the acceleration a. According to Newton's Second Law a force F is required to cause this acceleration:

$$F = ma. \tag{3.6}$$

This force is delivered to the mass via the coupling that connects the mass to the vehicle. If this coupling is some form of elastic element, which is operating in its linear region, then the displacement is proportional to the acceleration. Measuring this displacement therefore provides an estimate of the acceleration. A miniature cantilever made in silicon is a simple structure that has been used as the basis for such a device.

If the elastic properties of the material remain constant then the mathematical model of the system is a linear equation and the general relationships described in section 2.2 are appropriate. The damping ratio ξ needs to be approximately 1 so that the system does not oscillate and that the maximum deflection is reached speedily. This can be achieved by appropriate sensor geometry and utilizing the viscosity of the surrounding fluid (air). The solution to equation (2.1) for a step input then becomes

$$x = [1 - (1 + \omega_n t) e^{-\omega_n t}] \frac{l^3}{3EI} F \tag{3.7}$$

where E is Young's modulus, l the length of the beam and I the second moment of area $bd^3/12$, where b and d are the beam width and thickness respectively. For a cantilever it can be shown that

$$\omega_n = \sqrt{\frac{3EI}{ml^3}} = \frac{1}{2}\sqrt{\frac{Ebd^3}{ml^3}} \tag{3.8}$$

where m is the effective mass which can be taken as that of the end block if this is greater than that of the beam by a factor of about 30.

In addition to the simple cantilever, other designs have been developed and include the hammock, crab-leg and folded flexures (Gardner *et al* 2001). The methods developed to detect this displacement are described below, the most popular being the piezoresistance and capacitance types.

Piezoresistive

Accelerometers employing piezoresistance have been commercially available from the mid 1980s. Early designs were adaptations of the pressure gauge described above with an added mass in the centre of the diaphragm. In more recent designs the micro-inertial mass is suspended from struts(s) of material containing the piezoresistive elements. If silicon is used the 'spring' is normally nonlinear and so careful design of the geometry is essential if a linear output is desired. As described above, piezoresistive devices are prone to contamination and suffer from temperature drift. Encapsulation is needed and temperature compensation can be achieved using a dummy reference element.

As with the piezoresistance pressure gauge, using multiple sensors is a better alternative to having a dummy reference. This is easily illustrated in the simple cantilever. Sensors placed on the upper and lower surfaces suffer opposite stresses and when appropriately configured in a bridge network, as shown in figure 3.4, can multiply the sensitivity as well as compensating for temperature drift. $E_{\text{Thévenin}}$ is the open circuit potential difference.

Capacitive

The underside of the inertial mass is coated with a thin metallic film which acts as one of the plates of a parallel plate capacitor. The substrate is similarly treated to form the other plate. The capacitance C of a parallel plate capacitor is given by

$$C = \frac{A\varepsilon_0\varepsilon_{\mathrm{r}}}{d} \tag{3.9}$$

where A is the plate area, ε_0 is the permittivity of free space, ε_{r} is the dielectric constant of the material between the plates, and d is the plate separation.

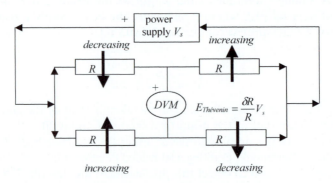

Figure 3.4. Deflection bridge with four active sensors.

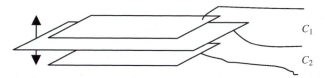

Figure 3.5. Differential capacitor system.

Variation in d is used to alter C. This is a nonlinear relationship which may be linearized using a bridge circuit or microprocessor (see Smart sensors, section 5.5). The capacitance is very temperature sensitive due to thermal expansion affecting d. This may be reduced by careful design incorporating the differential twin capacitor technique shown in figure 3.5. A further advantage of this differential system is that it can be incorporated into two arms of an a.c. bridge which linearizes the output signal, $E_{\text{Thévenin}} = (V_s/2d)x$, where $2d$ is the fixed separation of the outer plates and x is the displacement of the centre plate.

These devices became commercially available shortly after the piezo-resistive types. Reidemeister and Johnson (1995) report an application in automobiles. Thermal stability was achieved by matching the coefficients of thermal expansion of the beam, support base-pin and substrate in this single capacitor design. A guard electrode was also used to improve performance. The structure was enhanced using finite element analysis so as to measure low values of acceleration.

Analog Devices produced one of the first commercial capacitance-based devices described in more detail below.

Piezoelectric

Unfortunately silicon has no piezoelectric effect. Consequently, silicon-based sensors need to be coated in some piezoelectric thin film such as zinc oxide. The advantages of piezoelectric sensing are a wider working temperature range and smaller size. However, a significant disadvantage is charge leakage. This has the effect that the measurement of a steady acceleration appears to be zero after a few seconds. This restricts their application to vibration measurement. They are also more expensive than the above.

Resonance

A further design of accelerometer is based on the resonator technique. This is a similar concept to that described above for pressure measurement. Rather than the deflection of the inertial mass being measured, the change in the modulus of elasticity of the strut material due to varying stress is used. Changing the elasticity has the effect of changing the natural frequency of the micro-flexing structure. Piezoelectric drive and sense elements are used in a feedback loop to maintain resonance. The resonant frequency is the output signal indicating the acceleration.

Force-balance

Designs have become established using the force-balance technique. The concept is to use a magnetic or electric field to counter the acceleration force so that the inertial mass suffers virtually no displacement. Consequently, a linear response is achieved regardless of the geometrical structure of the mass suspension system. The output signal is derived from the feedback signal that generates the magnetic or electric field.

A commercial device using this is the ADXL-50 from Analog Devices, first available in 1991 and able to measure accelerations of 50g (Patel 1992). It is a surface-machined device and uses capacitance to monitor the movement. The fixed capacitor plates are a double comb-like structure mounted 1 μm above the surface on either side of the shuttle section. The shuttle has projections extending from either side. These form the moving capacitor plates which mesh with the combs. The total capacitance of the whole double-comb structure is only 1 pF and is less than 1 mm^2. It is configured to operate in differential mode as described above. The remainder of the chip contains an amplifier, oscillator demodulator and filter which provide a d.c. voltage directly proportional to acceleration. The complete chip is some 9 mm^2. Rather than allow the shuttle to move significantly a force-balance approach is adopted. A voltage is applied to the capacitor plates so that an electrostatic force is produced which opposes the acceleration force constraining the shuttle movement. This has the added advantage that the device can be tested without applying an acceleration.

3.2.4 Flow sensors

Thermal

The transfer of heat from a heated element to a flowing fluid was first proposed by C C Thomas in 1911 as a means of measuring flow. A simple implementation is in the so-called hot-wire anemometer used to measure wind speeds. Two self-heated sensors are incorporated in a deflection bridge network. One is shielded from the wind so as to act as a reference. In this way compensation is possible for changes in the ambient temperature of the wind.

For fluid flow in pipes, using non-miniature sensors and for microsensor implementation, a more robust approach is to have up-stream and down-stream sensing elements separated by a heater. Microsensors have been developed where the whole system is made on one silicon substrate.

Johnson and Higashi (1987) produced a device for gas flow measurement where two identical resistance thermometers were mounted on two micro-bridges above the substrate. The thermal conductivity of silicon is too high and would allow heat energy to drain into the substrate and so raise the power consumption of the device. Silicon nitride has lower conductivity and so was used to construct the bridges. A heating element was mounted between the two resistors, half on each microbridge, and gas flowed in a channel underneath. The tiny size of this microbridge structure resulted in low power consumption (typically 10 mW raised the temperature by 100 °C) and a fast response of about 3 ms. Various refinements have been reported measuring flow velocities of up to 30 m s^{-1}. Devices are available commercially from Honeywell for air mass flow sensing. Although in principle the technique should be applicable to liquids the risk of contamination is a significant problem. Any deposit on the surfaces would change the thermal conductivity and upset the calibration.

Resonance

Bouwstra *et al* (1990) have reported a silicon nitride, microbridge structure measuring $600 \times 200 \times 21$ μm. The resonant frequency of the structure was 85 kHz. Excitation was achieved thermally using thin film resistors. The elevated temperature was 20 °C above ambient. Piezoresistive sensors on the structure provided the output signal and a frequency shift of 800 Hz was observed for flows of 0–10 ml min^{-1}.

The advantages of resonant systems, as mentioned in chapter 2, are their high sensitivity, fast response and good stability. The major drawback in this device is the potential change in mass if contaminated. Only 'clean' fluids may be used.

Capacitance

Cho and Wise (1993) reported a gas flow sensor based on differential pressure. The design geometry of the dome structure was such that the mass flow rate was proportional to the differential pressure between inlet and outlet pipes. This differential pressure was used to vary the plate separation of an internally-mounted silicon micro-capacitor. Typical values of capacitance were 5–15 pF with 1 fF resolution measured using a CMOS switched capacitance circuit. It was reported that 1 fF corresponded to 0.13 Pa differential pressure.

The disadvantages of these devices, as of capacitance sensors generally, are their temperature drift and leakage current. Improved designs may reduce these.

The application of micro-flowmeters is set to increase over the years ahead due to the growing importance of micro-fluidics in MEMS (micro-electrical-mechanical systems) (Gardener *et al* 2001).

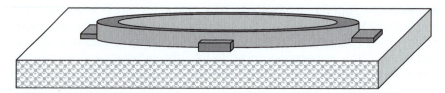

Figure 3.6. Miniature angular rate sensor.

3.2.5 Angular rate sensors

In section 2.7 the rate gyro was introduced as an angular rate sensor employing the Coriolis effect (see section 2.6). Semiconductor designs are mostly either some form of driven tuning-fork structure or a resonating cylinder or ring.

A tuning fork-type design was reported by Bernstein *et al* (1993). This was a micro-machined device that used a comb-drive structure. Microsensors of this type have evolved from the miniature tuning fork rate-gyro described in section 2.8. Whereas, the operating principle of the cylinder/ring devices is that of the vibrating tube rate-gyro described in section 2.7.

Johnson *et al* (1995) have reported the development of a ring-type, surface micro-machined, rate gyro. A metal ring was fabricated on a p-type silicon substrate by electroplating and photoresist technologies. A total of 32 electrodes were formed on the substrate around the ring (shown diagrammatically in figure 3.6) for sense and drive purposes. The substrate also contained CMOS circuitry for buffering and amplification. A second integrated circuit contained control loops, programmable DACs to correct for offset and gain errors, and a reference voltage source used to electrostatically polarize the ring.

There is a significant and expanding market for angular rate sensors. A number of manufacturers now produce rate-gyros, many of which have a similar external size of about 25 mm × 25 mm in section and about 18 mm high. However, there are a variety of constructional styles and performances. Those described above can be classed as truly solid-state devices. However, some are hybrid structures the primary sensor being a discrete ring, cylinder or, in one design an equilateral Elinvar prism, which is packaged with an integrated circuit.

3.3 Surface acoustic wave sensors (SAWS)

In section 2.10 sensors were described using a quartz crystal made to resonate mechanically throughout its bulk. Such oscillations are described as thickness-shear-mode (TSM). Surface acoustic waves (originally known as Rayleigh waves) are also mechanical waves but as the name suggests only propagate through the surface of the device. The disturbance is greatest at the surface but declines with depth and effectively only penetrates about a wavelength. Although the velocity of the waves is similar to that in TSM devices, being in the range $3.8–4.2 \times 10^2 \, \mathrm{m \, s^{-1}}$, frequencies are much higher. Devices can be made to operate at up to 5 GHz. Since the sensing ability is usually dependent on the fractional change in frequency $\Delta f / f$, such high frequencies are clearly an advantage. A further advantage of operating at microwave frequencies is that remote sensing by wireless operation is possible. The tiny sensor with antennae can be powered by the incoming electromagnetic wave and the returned wave transmits the data, see Varadan in Gardener *et al* (2001).

The operating principle of SAW sensors is that the measurand alters the velocity of wave propagation which in turn causes a change in the time interval between the injected and recovered signals. Devices typically have an area of a few mm^2. The piezoelectric effect is used for driving and sensing. Construction may be of a homogenous design. The drive and sense components and substrate are the same piezoelectric material. Alternatively, a semiconductor substrate can be used such as Si or GaAs. Since these semiconductors are not piezoelectric a coating of piezoelectric material is needed to generate the waves and sense their arrival. The most commonly used coatings are ZnO, AlN or $Pb(Zr,Ti)O_3$ (PZT) at about 5×10^{-7} m thick (0.05λ) for best ZnO–SiO$_2$–Si coupling. Using a semiconductor substrate means that signal conditioning circuitry may be built on to the device using monolithic fabrication technology. This is clearly an attractive option. However, the physics of the wave propagation changes. The wave velocity may no longer be independent of frequency. Also, the coating processes such as vapour deposition or sputtering mean that the piezoelectric layer will be polycrystalline and rough, which tends to disrupt wave coupling into the substrate.

The electrode pattern on the surface of the piezoelectric material is commonly the 'inter-digital' style, a comb-like structure, with up to 20 electrodes per set. The electrode patterns are fabricated using photolithography. The spacing between electrodes of the same polarity is one acoustic wavelength and is about 10 μm. The surface of the device is constructed as either a delay line or as a standing wave resonator. The structures of the two types are shown below.

Resonator saw sensor

The resonator SAW sensor (figure 3.7) has an echelle-type grating at each end of the active surface. These have the effect of reflecting the wave back towards the centre of the device so as to develop a standing wave. The sense and drive electrodes together with an amplifier, having automatic gain control, provide a positive feedback loop so as to maintain the resonant condition.

Changes in the propagation time between the transmit and receive electrodes will alter the resonant frequency, ω_n. The resonant frequency is

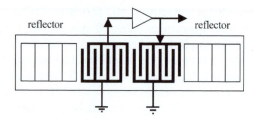

Figure 3.7. Surface acoustic wave sensor: resonator type.

given by

$$\omega_n = \frac{2\pi n - \varphi}{\tau} \tag{3.10}$$

where φ is the amplifier's phase shift, n is the excited mode number, and τ is the time delay, which is dependent on the measurand.

Delay line saw sensor

In this type of SAW sensor (figure 3.8) the oscillating frequency is fixed and the phase shift caused by the delay is measured. The delay line type lends itself well to a dual format enabling a reference device to be constructed on the adjacent area of the same substrate. Of the two operating types it is the more commonly used when a reference is required.

The phase shift ϕ between the transmitted and received signals caused by the propagation time of the SAW is given by

$$\phi = 2\pi f_0 t \tag{3.11}$$

where t is the time to travel over the surface between the electrodes. This can be expressed as

$$\phi = 2\pi f_0 \frac{l}{v} \tag{3.12}$$

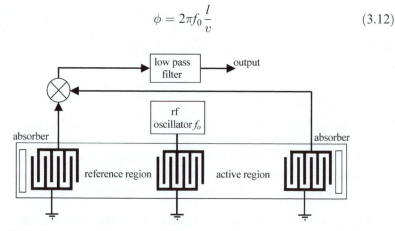

Figure 3.8. Surface acoustic wave sensor: delay line technique.

where l is the path length and v is the variable velocity altered by the measurand. For a dual delay line device f_0 is fixed. The difference in the phases of the two paths is given by

$$\phi_{\text{sense}} - \phi_{\text{reference}} = 2\pi f_0 \left[\frac{l_{\text{sense}}}{v_{\text{sense}}} - \frac{l_{\text{reference}}}{v_{\text{reference}}} \right] = \Delta\phi. \qquad (3.13)$$

An analogue multiplier can be conveniently used to combine the outputs of the reference and sense lines. Using the trigonometrical identity

$$\sin A \sin B = \tfrac{1}{2}\cos(A - B) - \tfrac{1}{2}\cos(A + B) \qquad (3.14)$$

gives

$$\sin(2\pi f_0 t + \Delta\phi) \sin 2\pi f_0 t = \tfrac{1}{2}\cos(\Delta\phi) - \tfrac{1}{2}\cos(4\pi f_0 t + \Delta\phi). \qquad (3.15)$$

A low-pass filter leaves just the $\tfrac{1}{2}\cos(\Delta\phi)$ term as the output. Some delay line sensors operate singly, without a reference section. In this type the r.f. oscillator signal can be multiplied with the received signal from the delay line to give the $\tfrac{1}{2}\cos(\Delta\phi)$ term.

3.3.1 Mass and gas/vapour sensors

A simple mass sensor, or microbalance, can be made using the SAW principle. Changes in mass, caused by some evaporation process for example, alters the wave velocity which is then output as a variable frequency. The change in frequency is given by

$$\Delta f = (k_1 + k_2) f_0^2 \Delta\mu \qquad (3.16)$$

where μ is the mass per unit area and k_1, k_2 are constants of the material. For Y-cut quartz k_1 is -9.8×10^{-8} and k_2 is $-4.18 \times 10^{-8}\,\text{m}^2\,\text{s/kg}$. Measurements as low as 1 picogram have been observed.

The design concept for a gas sensor is to have a coating that will selectively adsorb the molecules of the gas to be detected. This sorption changes the properties of the interface between coating and substrate which results in a change in the propagation time. This effect has been used in a resonator sensor to generate a change in the operating frequency. Wohltjers and Dessy reported the first SAW gas sensor in 1979. Research and development in this field has intensified. The advantages of SAW gas sensors compared with the bulk acoustic wave devices are

- the two orders of magnitude higher operating frequency means greater sensitivity,
- mounting the back of the planar substrate to a rigid surface gives greater stability than is possible with bulk resonators,
- miniature size means several can be built on one chip for multi-sensing and temperature compensation.

The design effort is now concentrated on the development of the chemical coating needed to detect the desired species. Vapour sensors have also been developed using a similar approach although here it is often the change in mass that changes the resonant frequency.

Sensors have been reported for measuring vapour levels of explosives, drugs, water and of the gases SO_2, H_2, NH_3, H_2S, NO_2, CO_2 and CH_4.

3.3.2 Temperature sensor

A $LiNbO_3$ substrate SAW temperature sensor has been reported by Neumeister *et al* (1990). This is a resonator-type device where the oscillation frequency ω_n is a function of temperature. This is due to the thermal expansion of the substrate causing an increase in the SAW propagation time. The temperature dependence is given by

$$\frac{\Delta\omega}{\omega_n} \approx (\alpha_1 - \alpha_2)\Delta T \qquad (3.17)$$

where α_1, α_2 are the thermal coefficients of expansion and velocity respectively. The device operates at about 40 MHz between $-50\,°C$ and $+150\,°C$ with a constant sensitivity of 4 Hz/°C. Resolution is high at 0.001 °C but repeatability is $\pm0.2\,°C$.

3.3.3 Strain sensor

The SAW delay line technique can be used to detect strain. If the propagation speed of the SAW wave is not affected by strain in the material then the resulting change in phase is only due to the change in the physical path length caused by the strain. Adapting equation (3.12),

$$\frac{\Delta\phi}{\phi} = \frac{\Delta l}{l}. \qquad (3.18)$$

But the tensile strain ε is defined as $\Delta l/l$ so $\Delta\phi$ is proportional to ε.

Placing the strain sensor on a miniature cantilever or similar structure enables the construction of an accelerometer. Similarly, a miniature pressure sensor can be constructed using a quartz diaphragm. The strain sensor is placed on the non-contact side of the diaphragm so that fluid cannot reach the SAW region since this would also cause a change in the frequency.

3.3.4 Magnetic sensor

The magneto-elastic effect, or magnetostriction, was observed by Joule in 1842. Magnetoelastic materials change their length when subjected to a longitudinal magnetic field. The effect is small, typically 10 ppm in moderate magnetic fields.

If such a magnetoelastic thin film is placed on a SAW device any magnetic fields present will cause stress to occur across the boundary layer due to the magnetoelastic material changing its length. This stress will alter the velocity of the SAW.

Hanna (1987) has reported a resonator-type SAW magnetic sensor. The relationship between resonant frequency and magnetic field was nonlinear but high sensitivity was achieved.

3.3.5   Conclusion

A number of SAW sensors have been developed and this is set to continue. To summarize, the attractive features of these devices are

- their high resolution,
- the facility to provide on-board electronics,
- the potential for wireless operation,
- the potential for high volume production using monolithic technology.

Their disadvantages are

- poor long-term stability,
- high temperature and humidity sensitivities.

Compensation techniques may help towards overcoming some of these.

3.4   Chemical microsensors

Traditionally, solutions have been analysed using glass electrodes for pH and ion-selective electrodes (ISEs) for ions such as K^+, Ca^{2+}, Cl^- and Br^-. Enzyme sensors and biosensors are also available commercially. However, during the research and development of semiconductor materials over the past 40 years it was noticed that surface properties were affected by a variety of impurities. In Japan in the early 1960s semiconductors were first used to detect gases in the atmosphere and later for the analysis of solutions. A great deal of effort has been put into the development of such devices and is continuing.

In recent years the world-wide growth in sales of chemical sensors has not suffered to the same extent as other areas of the global economy. The environmental and medical fields are expected to show the largest increase. Because of their limited initial use, chemical microsensors have greater predicted annual growth than the traditional sensors. Estimates of chemical microsensor growth rates are put at between 9.5 and 10.5% per annum for the years to 2006.

Although some developments have involved diode structures such as the Schottky, GCD (gate controlled device) and ICD (ion controlled device), the majority are based on the field effect transistor (FET). The functional part of the FET is the gate region. FET sensors can be classified according to the type of species to be detected. The following mnemonics have been coined:

ISFET	ion-sensitive FET
REFET	reference FET
CHEMFET	chemically sensitive FET
ENFET	enzyme FET
IMFET	an ISFET with an immunological reagent
BIOFET	an ISFET with a biologically active surface coating.

Most commonly, inorganic devices are referred to as ISFETs with their field of application specified. To appreciate their operating principles it is helpful to know a little chemistry.

3.4.1   Ions in solution

When an electrode is placed in an ionic solution a reaction takes place between the two so that a resulting equilibrium potential difference is produced. This potential difference between electrode and solution depends mainly, but not entirely, on the activity a of a particular ionic species. (In some situations the presence of a similar ion may modify the behaviour of the desired electrode–ion interaction.) The activity a of an ion in solution depends on the concentration c,

$$a = Kc. \tag{3.19}$$

However, K is not constant but is a function of c and the concentrations of other ions present. At low concentrations, $<10^{-4}$ mole litre^{-1}, $K \cong 1$. The modified Nernst equation relates the Thévenin emf to a,

$$E_{\text{Thévenin}} = E_0 + \frac{RT}{zF}\ln(a_x + ka_y) \tag{3.20}$$

where E_0 is a constant depending on electrode composition, R is the universal gas constant ($8.314\,\text{J K}^{-1}$), z is the number of charges per ion (1, 2 or 3), T is the absolute temperature, F is the Faraday number (96 493 Coulombs), a_x is the activity of the measurand ion x, a_y is the activity of ion y if present, and k is the selectivity of the electrode to y; $0 \leq k \leq 1$.

$E_{\text{Thévenin}}$ is the potential difference produced between the single active electrode and the solution in which it is immersed. Another electrode also has to be used in the solution in order to complete the circuit. The emf produced by this second electrode must be constant irrespective of the solution under test. It behaves in effect as a reference electrode delivering a fixed potential difference to the circuit.

3.4.2 Measurement of pH

One of the ionic activities most frequently measured, particularly by the process industries and when monitoring the environment, is that of the hydrogen ion. This is a measure of the acidity/alkalinity of the solution. The defining equation for pH is

$$pH = -\log_{10} a_H \tag{3.21}$$

where a_H is the activity of the hydrogen ion. In pure water dissociation of the water molecule occurs to a small extent,

$$H_2O \Leftrightarrow H^+ + OH^- \tag{3.22}$$

resulting in a hydrogen ion concentration of 10^{-7} moles per litre at $25\,^{\circ}C$. If the temperature rises, more dissociation occurs, increasing the hydrogen ion concentration. This would give the impression that the solution had become more acidic. However, as long as the H^+ and OH^- ions remain in equal concentrations the solution stays neutral. Consequently, for the pH meter to output the value 7, indicating neutrality, a temperature-dependent correction must be applied. Generally, therefore, pH meters have some form of temperature compensation system. Table 3.1 shows the pH values for three types of ionic solution.

The pH of a solution can be determined using electrodes and applying the Nernst equation. Assuming the sensing electrode is highly selective and only responds to a_H, equation (3.20) can be written as,

$$E_{\text{Thévenin}} = E_0 + \frac{RT}{zF} \ln a_H \qquad \text{for the hydrogen ion } z \text{ is } 1 \tag{3.23}$$

$$E_{\text{Thévenin}} = E_0 + \frac{RT}{F} \ln a_H \tag{3.24}$$

$$E_{\text{Thévenin}} = E_0 + \frac{RT}{F} \ln 10^{-pH} \tag{3.25}$$

$$E_{\text{Thévenin}} = E_0 - 2.303 \frac{RT}{F} pH \tag{3.26}$$

which at $25\,^{\circ}C$ gives

$$E_{\text{Thévenin}} = E_0 - 0.0592\, pH \text{ volts} \tag{3.27}$$

Table 3.1. Hydrogen ion concentration and pH values.

Ionic solution	Hydrogen ion concentration (moles per litre)	pH
Pure water at $25\,^{\circ}C$	10^{-7}	7
Strongly acidic	Tends to 10^0	Tends to 0
Strongly alkaline	Tends to 10^{-14}	Tends to 14

Equation (3.27) has been in use for a number of years for measuring pH. The traditional system uses a sensing electrode housed in a permeable glass membrane. As described above, to measure the potential difference between electrode and solution a further electrode is required whose $E_{Thévenin}$ does not change with the constitution of the solution. The output of this reference electrode E_{ref} is included in the calibration. These glass electrode instruments generally work well once conditioned but are bulky, fragile and expensive.

A new type of pH microsensor is now commercially available based on the MOSFET. Similar devices are also available for the measurement of other ions in solution, the family of sensors are the ion specific FETs or ISFETs.

3.4.3   The ISFET

The ISFET was the first attempt at using FETs as sensors and was initially proposed to measure pH by Bergveld in 1970. To help appreciate the operation of the ISFET it is useful to consider the operation of the MOSFET from which it was developed.

The acronym MOS describes the material sandwich in the construction of the FET, metal-oxide-semiconductor (figure 3.9). The semiconductor may be p- or n-type doped, to produce a PMOS or NMOS device. NMOS devices, using n-channel enhancement, have a faster switching action due to the holes in the p-type substrate being more mobile. This is not usually of any significance in sensor technology but is important in high-frequency electronic devices. NMOS is also preferred in very large-scale integrated circuit devices (VLSI's), since packing densities can be higher.

The constructional details of the n-channel, enhancement mode, NMOS FET are summarized below.

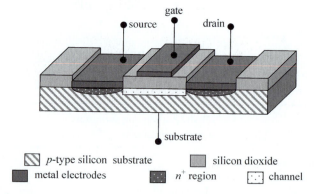

Figure 3.9. The NMOS FET.

- The p-type substrate is lightly doped so its resistance is high.
- The source-substrate n–p junction is of order $10^{10}\,\Omega$ unless forward biased.
- Normally the source and substrate connections shown on the diagram are connected internally.

When a positive potential is applied to the gate relative to the substrate, electrons are attracted to the surface of the p-type substrate providing a low resistance channel of order $1\,\text{k}\Omega$. The gate potential thus controls the current between drain and source. The resistance between gate and substrate is very high so that there is negligible gate current. The drain-source current I_{DS} is given by

$$I_{DS} = \frac{\mu_m C_0 w U_{DS}}{L}\left[U_{GSu} - U_t^+ - \tfrac{1}{2}U_{DS}\right] \tag{3.28}$$

where U_{GSu} is the potential difference between gate and substrate, U_t^+ is the threshold potential difference of the FET, μ_m is the effective mobility of electrons, C_0 is the capacitance per unit area of the gate structure, w is the width of the gate, L is the length of the gate, and U_{DS} is the potential difference applied between drain and source.

Figure 3.10 shows how the NMOSFET is adapted to form the ISFET.

Positive ions (cations) from the solution are absorbed into the porous membrane, which functions as the gate. These induce a negative charge surface layer on top of the p-type silicon thus providing a conducting

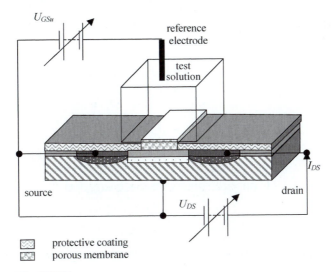

Figure 3.10. The ISFET structure.

channel between the source and the drain. As with the FET, the ISFET
substrate and source are at the same potential so that the n–p junction
does not become forward biased. As described above, the FET structure
can have a p- or n-type substrate and, depending on the type of ion, cation
(+) or anion (−), the current I_{DS} either increases or decreases with ion
concentration.

Initially it was thought that an ISFET could be constructed by simply
omitting the metal gate electrode. In this way the SiO_2 insulating layer
would absorb hydrogen ions that would in turn change the biasing of the
FET and so produce, effectively, a pH-sensitive device. However, using
SiO_2 alone, sensitivity was found to be only $30\,mV\,pH^{-1}$ and not sufficiently
stable when exposed to electrolytes. Consequently, a layer of silicon nitride
(Si_3N_4) is now commonly used on top of the SiO_2 to act as the absorbing
membrane. This device yields $50\,mV\,pH^{-1}$ linearly from 1.8 to 10 pH. The
sensitivity can vary depending on the thickness and the manufacturing
process. Other oxides, such as those of the more exotic metals, have been
examined as the active surface layer and some give $>50\,mV\,pH^{-1}$. Aluminium oxide, Al_2O_3, has also been used but gives a rather slower response
compared with the silicon compounds.

In the early 1970s it was thought that a reference electrode was not
essential. However, this is not the case. The reference electrode ensures the
device is biased at its optimum operating point.

The theoretical relationship for the drain-source current is obtained by
combining the Nernst equation with the FET equation

$$I_{DS} = \frac{\mu_m C_0 w U_{DS}}{L} \left[U_{GSu} - U_t^+ \pm \left(\frac{RT}{zF} \ln(a_x + ka_y) - E_{ref} \right) - \frac{1}{2} U_{DS} \right]$$

$$(3.29)$$

where E_{ref} is the potential difference of the reference electrode. A practical
configuration of the structure shown in figure 3.10 is depicted in figure 3.11.

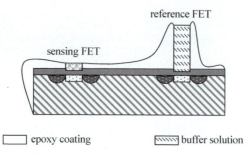

Figure 3.11. A practical ISFET structure.

The depositing of different electro-chemically active materials on top of the silicon nitride enables other ions to be detected. For example, sodium alumino-silicate produces a sensor for detecting Na^+ ions.

3.4.4 FET gas sensors

MOSFETs having a palladium or platinum coated gate can be used to construct a hydrogen gas sensor. Palladium causes the hydrogen molecule, H_2, to dissociate on its surface. These atoms then diffuse through the palladium and at the metal–insulator boundary become polarized. This provides an electric field that shifts the threshold voltage of the FET. Unfortunately, the rate of diffusion is slow and an integral heater is often employed to increase this. The metallic layer is operated at temperatures up to $150\,°C$.

Devices for sensing more complex species are continuing to be developed based on the construction of sophisticated adaptations to the pH and hydrogen ISFETs. For example, the CO_2 sensor shown in figure 3.12 is a pH sensor coated with a membrane which is permeable to CO_2. A small cavity dissolves the diffused CO_2 to form the weak carbonic acid, H_2CO_3, and the resulting acidic gel is measured for pH to determine the quantity of acid.

By the judicious choice of materials, their thicknesses, the metal gate structure and even including measurement of the operating temperature, gas sensors have been constructed for

hydrogen	H
ammonia	NH_3
hydrogen sulphide	H_2S
carbon dioxide	CO_2
hydrocarbons	C_xH_y where $y = 2x + 2$
alcohol vapours	$C_xH_{y-z}(OH)_z$

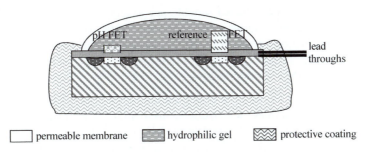

permeable membrane   hydrophilic gel   protective coating

Figure 3.12. The carbon dioxide ISFET sensor.

3.4.5 Biosensors

The design strategy is to take an established ISFET and surround it with a selective membrane that allows through the desired species. The production of these highly selective membranes is by the Langmuir–Blodgett technique. First a complex organic or polymer molecule is developed that is permeable to the species to be detected and has attached hydrophobic and hydrophilic radicals. These molecules are floated on pure water as mono-molecular layers. The hydrophilic–hydrophobic radicals ensure that the molecules are correctly orientated as they float. The FET is dipped into the surface and a single molecular layer is deposited across its surface as it is withdrawn. To aid the process a moving boom is used to maintain constant surface tension so that gaps or overlaps in the coating do not occur. Multiple layers can be deposited and may even be of different materials or orientations. In this way a highly selective molecular sieve is produced.

This membrane forms a small chamber over the ISFET. The chamber contains a reagent chosen to react with the biological species to be measured so as to produce a specific by-product. The by-product, which might for example be NH_3, CO_2 or a change in pH, is then detected by the ISFET. Bio-sensors of this type have been developed for measuring urea, glucose and penicillin. It is interesting to note that biosensors are far more selective than ordinary chemical sensors. This is entirely due to the superiority of natural systems over the engineered. Unfortunately, the typical life spans of such sensors are measured only in days (an ENFET for penicillin has been produced which remains active for up to 2 months).

An exciting application of the biosensor is its use *in vivo*. However, a design consideration when sensors are used within the blood stream is that proteins may affect the specification of the device. A further, more serious, concern is that the presence of the sensor may also encourage the formation of thromboses.

3.4.6 Conclusion

A small number of FET based sensors are available commercially. This is certain to be the beginnings of a major industry, the limitations to the variety of sensors being that of designing appropriate membrane barriers with reasonable longevity. The advantages of FET sensors over the traditional glass ISEs are as follows.

- Their low output impedance simplifies the signal conditioning.
- Their small size enables localized measurement, the use of small amounts of reagents, and use with catheters in medical applications.
- Their fabrication using well-established, semi-conductor industry, techniques enables multiple sensing on one chip for different species, the

reference electrode to be fabricated within the package, and temperature or light sensing on chip, for compensation.

- Faster response.
- Batch fabrication.

3.5   Optical sensors

3.5.1   The wave–particle duality

It is normal practice to classify electromagnetic radiation in terms of its wavelength. Light is the visible part of the electromagnetic spectrum, having wavelengths in the range 400–700 nm. The near infrared is taken as 0.7–1.7 µm, the main infrared as 1.7–30 µm and the far infrared as 30–1000 µm. Ultraviolet light has wavelengths in the range 10–400 nm. The near visible and the shorter wavelengths are sometimes referred to as 'soft' and 'hard' or vacuum ultraviolet respectively. However, depending on the circumstances, electromagnetic power may be considered as either the arrival of a continuous wave or of a stream of particles. This dichotomy applies to all particles whether they have mass or not and is known as the *wave–particle duality*.

Isaac Newton (1642–1727) was a believer in the particle or *corpuscle* theory of light, but diffraction and interference experiments, such as those of Thomas Young (1773–1829), reinforced the wave concept. However, in 1899 Lummer and Pringsheim conducted some experiments that could not be explained using a classical wave-model. The power spectral density of radiation from a perfectly emitting, or *black-body*, was measured at various temperatures (figure 3.13).

The theoretical studies of Wien (1896) and Rayleigh (1900), based on the classical statistical physics laws of Maxwell and Boltzmann, failed to completely account for these curves. It was Planck in 1901 who postulated that energy could only be absorbed/emitted by the black-body in discrete amounts he called *quanta*. He suggested that the transferred energy quanta, E, was determined by

$$E = hf \tag{3.30}$$

where h is Planck's constant and f is the frequency of the radiation. It was Einstein in 1907 who extended the concept further to say that energy in *any* system went in discrete intervals. He also gave the name *photons* to light quanta.

3.5.2   Definitions and units

It would, at first, seem sensible to use equations like the one above and Poynting's vector (see section 4.4.4),

$$P = E \wedge H \quad [\text{W m}^{-2}] \tag{3.31}$$

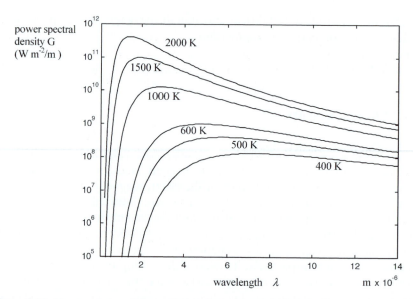

Figure 3.13. Power spectra of the radiation from a perfect black-body.

where E is the electric field strength in $V\,m^{-1}$ and H is the magnetic field strength in $A\,m^{-1}$, to define light units. However, the subject is complicated because the eye, like the ear, does not have a flat spectral response. Thus, for example, light from an argon ion (green) laser will appear to be brighter than light from a neon (red) discharge lamp of the same intensity. Consequently, a comparison between intensities is only meaningful if it is made in a restricted range of the spectrum. Historically, photometry has relied on standards based on a defined light source. No account is taken of the spectral details of the source: it is simply accepted as being spread throughout the visible spectrum in a standard way. The SI definitions and units for photometry given below have developed from these historical techniques.

The SI photometry units

Luminous intensity (I)—the candela (cd)

The candela is the luminous intensity perpendicular to the surface of a tiny, black-body emitter. The emitter is defined as having a surface area of $\frac{1}{6} \times 10^{-5}\,m^2$ and being at a temperature at which platinum freezes under a pressure of $101\,325\,N\,m^{-2}$ (this corresponds to a temperature of 2045 K).

Luminous flux (φ)—the lumen (lm)

Flux is an old English word meaning flow or discharge. The flow of light from a hypothetical point source is said to be isotropic, that is, it emits light uniformly in all directions. (Real light sources are not isotropic emitters.) The amount of light flowing in a given direction is the radiant power and is termed the luminous flux. The lumen is that flux flowing from the defined point source of one cd within a cone having a solid angle of one steradian (Ω). Generally,

$$\varphi = \int I \, \mathrm{d}\Omega \quad [\mathrm{lm}] \tag{3.32}$$

For an isotropic point source this would give the total flux as

$$\Phi_{\mathrm{total}} = 4\pi I \quad [\mathrm{lm}] \tag{3.33}$$

Illuminance—the lux (lx)

This is the irradiance, the amount of light falling on a surface. One lux of illumination is produced by one lumen falling on an area of one square metre, i.e.

$$1 \, \mathrm{lx} = 1 \, \mathrm{lm \, m}^{-2} \tag{3.34}$$

Typical values recommended for lighting installations are 500–600 lx for offices down to 100 lx for entrance porches when measured horizontally 1 m above floor level.

Luminance (cd m^{-2})

The brightness or radiance of a surface, either because it is self-luminous or illuminated, is termed its luminance. The SI unit of luminance is the candela per square metre.

The sensitivity of the eye

As mentioned above, the sensitivity of the eye is wavelength dependent and varies slightly between individuals. Photometric matching experiments have been carried out with numbers of subjects in order to relate the luminous flux perceived φ, to the actual spectral power distribution received φ_{p}. The relationship can be expressed as

$$\varphi = K_{\mathrm{m}} \int_0^\infty \varphi_{\mathrm{p}} V(\lambda) \, \mathrm{d}\lambda \tag{3.35}$$

where $V(\lambda)$ is the relative luminous efficiency of the eye and K_{m} is a constant. The Commission International de l'Eclairage (CIE) in 1971 recommended an agreed set of values for $V(\lambda)$. These values are sometimes referred to as the standard eye-ball. The form of $V(\lambda)$ is shown in figure 3.14.

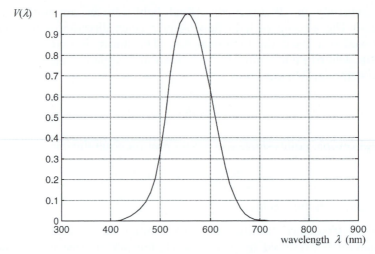

Figure 3.14. The agreed relative spectral response of the eye.

Practical photometry units

Researchers in the various optoelectronic fields tend not to refer to the SI photometry units because of the inconvenience of having to have standard sources and detectors of defined spectral response. Also, since much optical research is based on narrow spectral bands often in invisible parts of the spectrum, using photometry units is inappropriate. Consequently, researchers working at a particular narrow band of wavelengths will simply measure the electrical output of the detector. Then, knowing its active surface area and its efficiency at these wavelengths, the light intensity in $W\,m^{-2}$ can be determined. Researchers commonly use the following units which are based on the cm^2 as a unit area. This of course is not an SI unit.

Power spectral intensity of a source $S(\lambda)$

This is the power emitted per unit surface area of a source between λ and $\lambda + \delta\lambda$, into a cone of unit solid angle. In practice the units $W\,cm^{-2}\,\mu m^{-1}\,sr^{-1}$ are commonly used for $S(\lambda)$.

Brightness B

This is the total power emitted by a source over all wavelengths, per unit surface area, into a cone of unit solid angle:

$$B = \int_0^\infty S(\lambda)\,d\lambda \quad [W\,cm^{-2}\,sr^{-1}]. \tag{3.36}$$

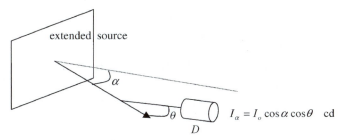

Figure 3.15. Lambert's cosine law.

Radiant flux R

This is the total power emitted from a source into a cone of unit solid angle between λ and $\lambda + \delta\lambda$

$$R = A_s S(\lambda) \quad [\mathrm{W}\,\mu\mathrm{m}^{-1}\,\mathrm{sr}^{-1}] \tag{3.37}$$

where $S(\lambda)$ is assumed constant over the surface area A_s of the source.

The diffuse source

An extended, diffuse source is one that behaves like an aggregate of point sources and so has an equal brightness when viewed from any angle. Such extended sources are sometimes said to be *Lambertian*. However, because of this aggregate of point sources, the luminous intensity is no longer equal in all directions. Instead a cosine dependence occurs and is termed *Lambert's cosine law*,

$$I_\alpha = I_0 \cos\alpha \quad [\mathrm{cd}]. \tag{3.38}$$

Also, if a detector D is placed at some angle θ to this direction, then the illumination it receives is further reduced by a factor of $\cos\theta$, as shown in figure 3.15.

3.5.3 Light sources

A large number of optical sensors are available each having a particular spectral sensitivity, speed of response, physical size and cost. Before examining light sensing devices it is worthwhile considering the types and nature of light emitters available. In this way an appropriate choice of sensor can be made.

Incandescent objects

Bodies such as the sun, filament lamps and molten metals give out a continuous spectrum of radiation similar to that of the perfect black-body

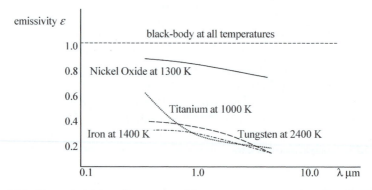

Figure 3.16. The emissivities of some materials.

(figure 3.13). However, the closeness of the emission to that of the black-body can be poor, especially at lower temperatures. Consequently, a correction factor called the emissivity ε can be defined, which is both wavelength and temperature dependent,

$$\varepsilon(\lambda, T) = \frac{\text{actual radiation}}{\text{black-body radiation}}. \tag{3.39}$$

The emissivities for some materials are illustrated in figure 3.16.

Filaments in filament lamps are usually made of tungsten. The lamp's glass or quartz envelope will filter the emission spectrum of the hot tungsten. Also, the tungsten slowly evaporates during its working life and forms a coating on the inside of the envelope. This coating also absorbs light. It has become common practice to add a small trace of a halogen, such as iodine, into high performance lamps. Tungsten–halogen compounds are volatile and re-evaporate from the glass surface whenever it is switched on. These materials also affect the transmitted spectrum.

The following relationships may be useful in determining ideal filament performance.

(a)   Planck's equation gives the power emitted per unit area of source, per unit wavelength from a black-body at an absolute temperature T,

$$G = \frac{C_1}{[\exp(C_2/(\lambda T)) - 1]\lambda^5} \quad [\text{W m}^{-2}/\text{m}] \tag{3.40}$$

where C_1 is 3.7418×10^{-16} W m^2 and C_2 is 1.4388×10^{-2} m K. This relationship generates the black-body power spectral density curves of figure 3.13.

The power, P, emitted per unit surface area from the whole spectrum is the integral of G over all wavelengths,

$$P = \int_0^\infty G \, d\lambda \tag{3.41}$$

which can be shown to give

$$P = \sigma T^4 \quad [\text{W m}^{-2}].$$ (3.42)

The term σT^4, is the Stefan–Boltzmann law for a black-body.

(b) The Stefan–Boltzmann law describes the radiated power, P, per unit area of emitter surface. The presence of another body, above absolute zero, near the radiator will cause the radiator to absorb power. The emissivity constant also applies to absorption by the non-black-body. The total, net, radiated power is

$$P = \varepsilon\sigma(T_B^4 - T_0^4) \quad [\text{W m}^{-2}]$$ (3.43)

where $\varepsilon(\lambda, T)$ is the emissivity, σ is the Stefan–Boltzmann constant $(5.670 \times 10^{-8}\,\text{W m}^{-2}\,\text{K}^{-4})$, T_B is the absolute temperature of the body, and T_0 is the surrounding absolute temperature.

(c) *Wien's displacement law* describes the shift of the spectral peak λ_p of the function G with the absolute temperature T,

$$\lambda_p = a/T$$ (3.44)

where a is a constant of value $2.892 \times 10^{-3}\,\text{m K}^{-1}$.

The black-body function G for some incandescent source can be related to the power spectral intensity function $S(\lambda)$. If the source is emitting isotropically into a hemisphere then

$$S = \frac{\varepsilon G}{2\pi} \times 10^{-10} \quad [\text{W cm}^{-2}\,\mu\text{m}^{-1}\,\text{sr}^{-1}].$$ (3.45)

Discharge lamps

When light is emitted from isolated ions, atoms or molecules, then line or band spectra are produced. These are precise characteristic frequencies and have little temperature dependence. Common examples are neon and sodium lamps. In a discharge lamp excitation of the gas is by collision with free electrons. Ions and atoms while excited have their outer electron(s) in higher orbitals. Molecules in addition to this can have rotational and vibrational energy. When the ions, atoms or molecules return to their ground states, quanta are released of characteristic wavelength. In high-pressure arc lamps a far greater number of excited states are possible, resulting in a spectral distribution graph rather than just line spectra. These distributions approach the black-body spectrum to a greater or lesser extent with the additional line spectra. The ultraviolet content is higher in high-pressure arc lamps than in the black-body spectrum. The exact nature of these spectral distribution graphs depends on the gas composition, its pressure, and the type of discharge and lamp geometry.

Light-emitting diodes (LEDs)

These are p–n junction diodes which, when forward biased, emit optical radiation. Holes and electrons in the junction region recombine, thus dropping from the conduction band to the valence band. The photon released corresponds to this energy reduction. The process is called *injection luminescence* since it is caused by the injection of charge carriers into the junction region when forward biased. Schockley derived the equations for an ideal diode: the current i_d is given by

$$i_d = i_s \left[\exp \left(\frac{eV_d}{kT} \right) - 1 \right] \tag{3.46}$$

where i_s is the reverse saturation current, e the charge of the electron, k Boltzmann's constant, T the absolute temperature and V_d the bias voltage. When forward biased, equation (3.34) approximates to

$$i_d = i_s \exp \left(\frac{eV_d}{kT} \right). \tag{3.47}$$

The luminous intensity is approximately proportional to the forward current. A number of devices are available constructed from combinations of aluminium (Al), gallium (Ga), arsenic (As), phosphorus (P) and indium (In), each having its own spectral distribution peaking at the wavelength λ_p:

GaP	$\lambda_p \cong 0.560\,\mu m$	green
GaAsP	$\lambda_p \cong 0.655\,\mu m$	red
GaAlAs	$\lambda_p \cong 0.8-0.9\,\mu m$	near infrared
InGaAsP	$\lambda_p \cong 1.3-1.6\,\mu m$	near infrared

There are two basic geometries giving rise to surface emitting and edge emitting types. The edge emitters give a much narrower beam. Turn on/off times are typically 10 ns giving modulation bandwidths of up to 10^8 Hz. Luminous intensities are of the order of 1 mcd. Infrared is preferred in optical fibre applications because of its better transmission through silica, compared with visible light. This is described further in section 4.2.2. Typical operating conditions have i_d set at about 20 mA, and V_d about 1.6 V; total light output is between 1 and 10 mW. Figure 3.17 shows the emission spectrum of a GaAlAs LED. The brightness B is the area under the curve and is approximately 10^2 W cm^{-2} sr^{-1}.

LASERs (Light Amplification by the Stimulated Emission of Radiation)

The lasing medium used in the first laser was a crystal of ruby in rod form. It was pulse-energized from a flash lamp. Lasers are now available with lasing media made of gas, dye-solution or semiconductor. The ions, atoms or molecules of the lasing medium have to undergo a *population inversion*. This means that they are put into an unstable high-energy state. They can

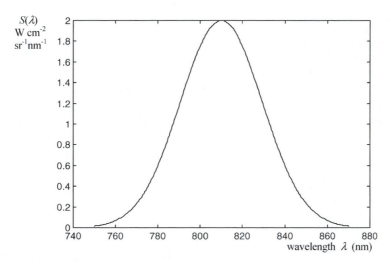

Figure 3.17. Spectral output of a GaAlAs LED.

spontaneously return to their ground state by emitting a photon. But they can be triggered or *stimulated* to do so by photons of the same type as to be emitted. Consequently, an avalanche effect can occur. Highly reflecting mirrors are mounted at each end of the lasing cavity to ensure that this inefficient process continues. Photons emerge in-phase and usually highly collimated providing an intense, monochromatic, coherent beam. In gas lasers bandwidths may be less than 10^{-5} nm. Semiconductor laser bandwidths are not as narrow. Also, the total light output may only be about the same as the ordinary LED due to the smaller emitting area. The brightness B in figure 3.18 is approximately 10^5 W cm^{-2} sr^{-1}.

It is important to remember that all LED devices are adversely affected by temperature. The intensity variation is

$$I = I_0 \exp(-\alpha\theta) \tag{3.48}$$

where α is the temperature coefficient of intensity, and θ the temperature in degrees Celsius. Also, the spectral peak λ_p shifts with temperature according to

$$\lambda_p(\theta) = \lambda_p(0) + \beta\theta \tag{3.49}$$

where $\lambda_p(0)$ is the spectral peak at $0\,°$C. In GaAsP devices, for example, α is 10^{-3} K^{-1} and β is 0.193 nm K^{-1}. In addition, laser LEDs suffer from changes to the optical cavity caused by temperature variation. This can result in changes to λ_p as well as *mode hopping*, a change of the standing wave pattern inside the resonant cavity.

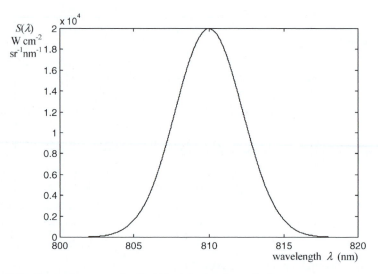

Figure 3.18. Spectral output of a GaAlAs injection laser diode.

3.5.4   Sensor types

When designing an optical measurement system the following criteria may need to be considered:

- sensing at the appropriate wavelength,
- inherent noise in the sensor,
- linearity,
- speed of response for the desired bandwidth,
- stability,
- compatibility with the overall system, particularly the optical geometry.

Photodetectors can be broadly classified into two groups: thermal detectors and quantum detectors. In thermal detectors the absorption of photons causes bulk warming of the sensor. These sensors have a very wide operating range of wavelengths, are relatively slow and have a low sensitivity (see section 3.6). In quantum detectors individual photons must deliver enough energy to overcome some threshold to release charge, the photoelectric effect. There are two groups of quantum detectors:

(a) external, or surface, photoelectric effect devices, the prime example of which is the photomultiplier, and

(b) internal photoelectric effect devices; the two types of semiconductor photodetector are the photoconductor and the junction photon detectors, the photodiode and the phototransistor.

(a) Surface photoelectric detectors

The photomultiplier

These are extremely sensitive devices used principally in research. They are vacuum tube structures (see figure 3.19) requiring a stable, high voltage d.c. supply of a few kV. A photon, passing through the glass window, strikes the cathode and may liberate an electron from the surface (photo-electric effect). There are a series of electrodes along the inside of the tube, called dynodes, at progressively higher voltages. The photo-electron accelerates towards the first dynode and on striking the surface has sufficient kinetic energy to liberate several more electrons. These then accelerate to the next dynode and the process repeats. Avalanche multiplication occurs. The resulting photocurrent is collected at the anode. The photo-active surface is made from exotic alloys of the reactive metals. The dynodes are also coated in similar materials. Photomultipliers are made with differing overall gains and spectral responses and are usually too sensitive to be operated in daylight. Some are sensitive enough to detect the arrival of a single photon. They have high linearity, very high bandwidth and low dark current. Their advantages when comparing with semiconductor devices are their ultraviolet response and large detecting area.

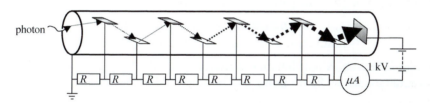

Figure 3.19. Schematic structure of a photomultiplier.

(b) Semiconductor photodetectors

These are devices in which a photon from the incident light beam causes an electron to be excited from the valence band to the conduction band. The detector will only respond to photons whose energy is

$$E_G > hf \quad \text{or} \quad hc/\lambda \tag{3.50}$$

where E_G is the energy gap between the valence and conduction bands. The two main types of detector are described below.

Photoconductive detectors

As the name suggests, these devices behave like variable resistors and are sometimes called light-dependent resistors. When the material is irradiated, if the incident photon has an energy $>E_G$, then an electron–hole pair is produced. The electron and hole drift to the appropriate electrodes under the influence of the applied electric field. The resulting photocurrent affects the conductivity of the material which magnifies the device current. The relationship between resistance R, and total incident light power is extremely nonlinear

$$\log_{10} R = a - b \log_{10} PA \qquad (3.51)$$

where P is the intensity, A is the sensing area and a, b are constants. The three most common devices and the wavelengths of their response peaks λ_p are

cadmium sulphide   CdS $\lambda_p \cong 0.53\,\mu\mathrm{m}$ visible

lead sulphide   PbS $\lambda_p \cong 3.0\,\mu\mathrm{m}$ infrared

indium sulphide   InSb $\lambda_p \cong 5.3\,\mu\mathrm{m}$ infrared.

Response times are long, typically 50 ms. Dark resistance is of order $1\,\mathrm{M}\Omega$ and falls nonlinearly with increase in illuminance to approximately $1\,\mathrm{k}\Omega$. Temperature stability is poor and depends on composition, fabrication method and light level. Also, the zero value drifts with time by up to 20% over 1000 hours. The application of these devices is therefore limited, since they are not suitable for fast response or precision measurement. They are, however, reasonably sensitive. CdS light-dependent resistors, for example, have a gain of 10^3. The gain is defined as the ratio of

$$\frac{\text{the number of electrons flowing}}{\text{number of electrons generated, per unit time}}$$

(holes have a short half-life and so can be ignored). They are used in optical switching systems, light alarms and photographic exposure meters.

Junction photon detectors

There are two classes of semiconductor junction device: the photodiode and the phototransistor. The photodiode is the most popular type with the PIN photodiode being the most widely used device.

Photodiodes

If the photon has sufficient energy, and if an interaction occurs, an electron–hole pair is produced. The p–n junction field sweeps these apart to the appropriate electrodes. The ideal diode equation for any p–n junction diode was given above and is repeated here:

$$i_d = i_s \left[\exp\left(\frac{eV_d}{kT} \right) - 1 \right]. \qquad (3.52)$$

The equivalent circuit for the photodiode is shown in figure 3.20; internal noise generation sources have been omitted. The photocurrent i_p created

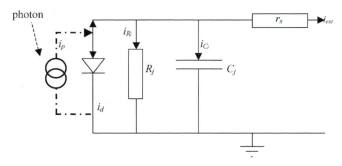

Figure 3.20. Photodiode equivalent circuit.

by the incoming photons can be expressed as

(number of photons arriving/sec) × (conversion efficiency)

× (charge of the electron)

This can be re-expressed in terms of the incident light power as

$$\frac{\text{(incident power)} \times \text{(conversion efficiency)} \times \text{(charge of the electron)}}{\text{energy per photon}}.$$

Expressing this algebraically gives

$$i_p = e\eta PA \frac{\lambda}{hc} \qquad (3.53)$$

where η is the quantum efficiency, the number of electrons collected per incident photon, e is the charge of the electron, P is the incident light intensity in W m^{-2}, A is the surface area in m^2, h is Planck's constant, and c is the speed of light.

Not all of the generated photocurrent i_p reaches the diode terminals. From the diagram of the diode equivalent circuit (figure 3.20), using Kirchhoff's current law gives

$$i_p = i_d + i_{R_j} + i_{C_j} + i_{ext}. \qquad (3.54)$$

Note that the forward biased terminal current is $-i_{ext}$ so that increasing the light intensity, which causes an increase in i_p, results in a reduction in the forward terminal current. The illumination has the effect of lowering the current–voltage characteristic as shown in figure 3.21.

The advantages of photodiodes when compared with photoconductive cells are their higher sensitivity, faster response time, smaller size, better stability and, when biased appropriately, excellent linearity.

A photodiode is made to operate at a particular point on its characteristic by the choice of biasing. Generally, diodes may be operated with or without an external bias voltage. A different performance results for each of the three possible biasing configurations. Biasing is usually achieved via

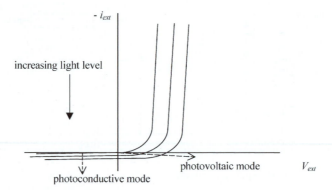

Figure 3.21. The photodiode characteristic.

the signal conditioning circuit. These are described below with their relative merits.

Photodiode signal conditioning circuits

Photovoltaic mode

Without an external bias the device is operating in the photovoltaic mode. Figure 3.22 shows a simple and buffered signal conditioning circuits.

The incident radiation generates a voltage across the junction, the electron–hole pairs being swept apart by local fields. The value of R_L is chosen to be very high, so that i_{ext} is practically zero. For low frequencies leakage through C_j will be negligible so that i_{C_j} can be assumed to be zero. Equation (3.54) then becomes

$$i_p = i_d + i_{R_j} \tag{3.55}$$

$$i_p = i_s \left[\exp\left(\frac{eV_d}{kT}\right) - 1 \right] + \frac{V_d}{R_j} \tag{3.56}$$

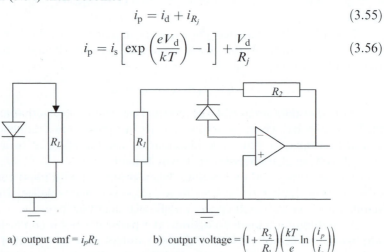

a) output emf $= i_p R_L$ b) output voltage $= \left(1 + \frac{R_2}{R_1}\right)\left(\frac{kT}{e}\ln\left(\frac{i_p}{i_s}\right)\right)$

Figure 3.22. Signal conditioning circuits for the photovoltaic mode.

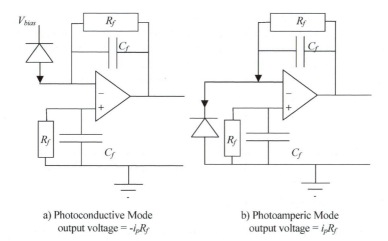

a) Photoconductive Mode
output voltage = $-i_p R_f$

b) Photoamperic Mode
output voltage = $i_p R_f$

Figure 3.23. Simple signal conditioning circuits for biased photodiodes.

and re-arranging gives

$$\exp\left(\frac{eV_d}{kT}\right) = 1 + \frac{i_p}{i_s} - \frac{V_d}{i_s R_j}. \tag{3.57}$$

Typically, the photocurrent will be of the order of $10\,\mu A$, whereas the reverse saturation current is of order $10^{-8}\,A$, so that $i_p \gg i_s$. Also, V_d will be approximately the same size as the product $i_s R_j$, so equation (3.57) can be written

$$\exp\left(\frac{eV_d}{kT}\right) = \frac{i_p}{i_s} \tag{3.58}$$

$$\frac{eV_d}{kT} = \ln\left(\frac{i_p}{i_s}\right) \tag{3.59}$$

$$V_d = \frac{kT}{e}\ln\left(\frac{e\eta PA\lambda}{i_s hc}\right) \tag{3.60}$$

where V_d is the output potential difference across R_L and has a logarithmic response with P and λ. Figure 3.21 gives the output relationship when an op-amp is used.

In the photovoltaic mode of operation the frequency response is only moderate but the noise is low and is mainly thermal noise, which is spectrally flat. This biasing approach is thus preferred for low noise applications at low frequencies. Below 100 kHz the photovoltaic mode provides a better signal to noise ratio than the photoconductive biasing mode described below and, at frequencies of less than 1 kHz, is far superior due to lack of $1/f$ noise. A disadvantage is that the output voltage varies logarithmically with light intensity as equation (3.60) shows. However, this may be put to good use, for example in the measurement of liquid depth. Absorption is an exponential relationship so having a logarithmic sensor response would give a linear output. A further most important application of diodes in the photovoltaic mode is as a solar cell. However, the physical structure of solar cells is different since large areas are needed and power transfer is required.

Photoconductive/photoamperic modes

With an external reverse bias the device is operating in the photoconductive mode. The junction field is increased, reducing charge carrier transit times and junction capacitance because of the increase in the width of the depletion region. Thus high-frequency performance is improved. Thermal noise and shot noise are present with $1/f$ noise becoming dominant below 1 kHz.

If a bias of zero volts is applied, this is referred to as the photoamperic mode. This mode has very low photodiode leakage current but is slower than photoconductive. Photodiode sensitivity is linear in both of these modes.

For reverse bias operation the diode current i_d is the leakage current i_s, and is typically 10^{-8} A. (If the bias is zero then i_d is zero.) The reverse bias voltage applied will be of the order of 10 V and, since r_s is small, this will be the same as V_d. R_j is typically in the region of $10^8 \, \Omega$. Consequently, i_{R_j}, which is V_d/R_j, will be approximately 10^{-7} A. The current i_{C_j} will be small because of the decreased capacitance so to a good approximation equation (3.54) becomes

$$i_p = 10^{-8} + 10^{-7} + i_{ext}. \tag{3.61}$$

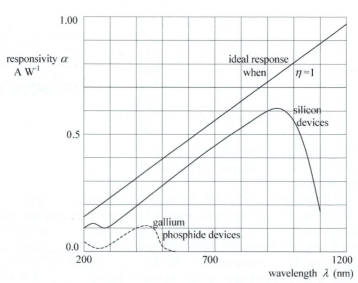

Figure 3.24. Spectral response of photodiodes when operated in the photoconductive mode.

Thus, for a typical photocurrent of $10\,\mu A$, the external current will effectively be equal to the photocurrent,

$$i_{\text{ext}} = e\eta P A \frac{\lambda}{hc} \tag{3.62}$$

which shows a linear response with P and λ and is sometimes written as

$$i_{\text{ext}} = \alpha P A \tag{3.63}$$

where $\alpha = (e\eta\lambda)/hc$ and is termed the responsivity with units of $A\,W^{-1}$. The spectral response is shown in figure 3.24.

There are four main types of photodiode. These are briefly described below.

The p–n junction photodiode

The descriptions in the previous paragraphs relate to the p–n junction diode. The semi conductor material may be of silicon or of GaAsP.

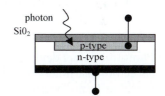

PIN photodiode

This diode has an intrinsic layer of silicon between the extrinsic p and n layers. This lowers the device capacitance and produces a much faster response than for the p–n junction photodiode, being typically $0.4\,\mu s$. Another reason for its popularity is that the intrinsic region can be tailored to optimize the desired quantum efficiency and bandwidth.

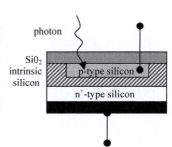

Schottky photodiode

An extremely thin metal film (of order $10^{-8}\,m$ thick) forms a Schottky barrier with an n-type semiconductor. The thin metal film enhances the response to the ultraviolet end of the spectrum. An anti-reflection coating is essential to reduce losses. With the right choice of metal and coating quantum efficiencies can reach 70% and response times in the region of 1 ns.

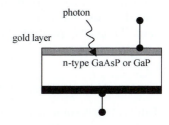

Avalanche photodiode

This device is operated under high reverse bias voltage. The liberated electron–hole pairs accelerate to such kinetic energies that, when collisions with other atoms occur, secondary charge carriers are produced. Avalanche photodiodes have ultra-fast response times and can operate up to GHz frequencies. The

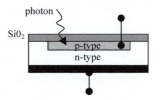

breakdown voltage for these devices when fabricated in silicon is in the region of 100 V. Photocurrents of up to 1 mA can occur when operated near this potential difference. Avalanche gain may be between 100 and 10 000 and can substantially reduce the noise equivalent power because the gain is achieved within the device and not subsequently in some signal conditioning circuit. This gives these photodiodes a significant advantage over non-avalanche types. However, the power supply voltage needs to be stable to ± 20 mV, in order to maintain constant gain.

Phototransistors

Phototransistors detect photons in the same way as photodiodes, the enlarged collector–base junction acting as the sensing region. For an n–p–n transistor, if the base is floating then the holes created forward bias the base–emitter junction. The bipolar-transistor, common-emitter mode, current-gain parameter h_{FE} amplifies the photocurrent. The overall current is $(1 + h_{FE})i_p$. Thus, phototransistors are more sensitive than p–n photodiodes, but their response is less linear. Also, the capacitance of the forward-biased, emitter junction causes a slow response of typically 1–5 μs and so frequency performance is poor. Also, because the collector current is temperature-sensitive, phototransistors are less temperature stable than photodiodes. Phototransistors manufactured as three-terminal devices can be biased and operated in the normal bipolar fashion giving greater flexibility in system design. For this reason the phototransistor may be a more appropriate optical sensor than the photodiode in some applications. They are often used in reading, writing and encoding systems. Photodarlington configurations are available. Here the phototransistor emitter current is cascaded into the base of a normal bipolar transistor, giving a multiplicative current gain. Typically, photodarlington devices are ten times more sensitive but ten times slower.

Commercially available devices

Table 3.2 lists some devices with an indication of their prices as advertised at the beginning of the year 2002. Price depends on a number of characteristics other than the few listed. These include the size of the detecting area,

Table 3.2. A selection of semiconductor photodetectors.

Device / manufacturer	Type	Spectral range	Response time	Cost (£)
NORP12	Photoconductor (CdS)	λ_p 530 nm	rise 18 ms fall 120 ms	1.51
VT935G / E G & G Vactec	Photoconductor	λ_p 550 nm	rise 35 ms fall 5 ms	0.67
SFH213 / Siemens	Si p–n	λ_p 850 nm	5 ns	0.47
AEPX65 / Centronic	Si p–n	λ_p 800 nm	1 ns	14.49
BPX65 / Centronic	Si PIN	λ_p 850 nm	3.5 ns	6.11
BPW34 / Siemens	Si PIN	λ_p 850 nm	20 ns	0.73
OP993 / Optek	Si PIN	λ_p 890 nm	5 ns	1.57
C30902E / EG & G	Si avalanche	400–100 nm	0.5 µs	120.36
SDP8405 / Honeywell	Phototransistor	λ_p 850 nm	15 µs	0.71
BPV11 / TFK	Phototransistor	λ_p 950 nm	3.8 µs	0.35
SFH309FA-4 / Infineon	Phototransistor	λ_p 900 nm	10 µs	0.42
BPW38 / Fairchild	Photodarlington	λ_p 850 nm	250 µs	1.30

efficiency, additional optical filters or lenses, the angle of acceptance, level of dark current and specific spectral response. All of these vary between manufacturers. Purchasers need to study individual data sheets carefully.

The charge-coupled device optical sensor

Charge-coupled devices (CCDs) are a family of silicon semiconductor devices capable of performing the general functions of analogue data handling and digital or analogue memory. The CCD concept was invented in 1969 at Bell Labs by George Smith and Willard Boyle whilst researching into new forms of computer memory. The combination of CCD with an optical sensing array was fabricated soon after to produce a very versatile optical sensor that has enjoyed rapid growth over the intervening years. The two discrete parts, the optical sensing process and the charge-coupling action, are described below.

Sensing

Sensing is achieved by the photoelectric effect. Free electrons are created at sites in the silicon illuminated by photons in the spectral range 300 to 1100 nm. The response peak is around 800 nm. The device is constructed having an array of these *photosites* which are typically 10 µm across. The number of free electrons generated in each cell is directly proportional to the intensity of the incident radiation on that cell. If the incident radiation intensity pattern on the array of cells is a focused image then the amounts of charge, or *charge-packets*, created in each cell will be a representation of the viewed object. Each charge-packet corresponds to a picture element or *pixel*. The array of contiguous cells may be one- or two-dimensional.

Charge coupling

Specific locations, called *storage elements*, are created in p-type silicon by asymmetric doping in conjunction with the field from a pair of *gate* electrodes that are close to the surface of the silicon. Figure 3.25 illustrates a

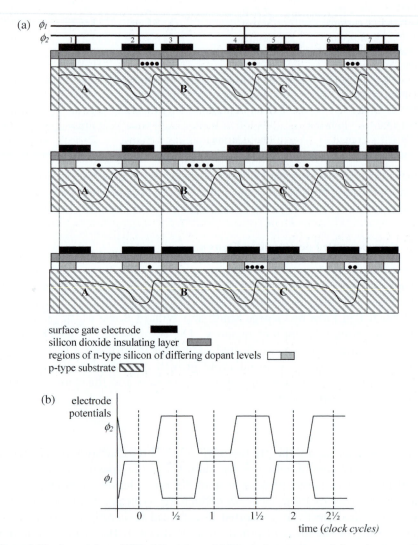

Figure 3.25. A two-phase CCD shift register. (a) Physical structure and changing potential well patterns. (b) Clock voltage waveforms.

two phase-type device. An insulating layer of silicon dioxide separates the electrodes and storage elements. The storage elements form a contiguous, linear array. The different concentrations of n-type dopant create the required asymmetric electric field.

The potentials on the gate electrodes can be alternately raised and lowered so as to cause the individual charge-packets beneath them to be passed from one storage element to the next. This is illustrated in figure 3.25(a) for the three cells A, B, C within the array. The two complementary clock voltage waveforms ϕ_1 and ϕ_2 (figure 3.25(b)) are connected to the alternate, closely spaced, gate electrodes on the surface of the thin insulating layer of silicon dioxide.

At the start of the cycle ϕ_1 voltage is HIGH and the charge packet of electrons in storage element A is trapped by the potential well under gate electrode 2. In the next half-cycle the potential well under gate 2 is reduced due to ϕ_1 having gone LOW. At the same time the adjacent electrode 3 connected to ϕ_2 has gone HIGH. Consequently, the electron charge packet is attracted to the new potential well under electrode 3. A further half-cycle later the potential well under electrode 3 is reduced due to ϕ_2 going LOW and the electron packet moves to the new well under electrode 4 which has gone HIGH.

The line of storage elements thus operates as an analogue shift register and, because the transfer of charge from one storage element to the next is very efficient, the amount of charge in each charge-packet stays substantially the same. Shift registers are available with up to 4096 contiguous elements.

The optical CCD device

Figure 3.26 illustrates the two processes of sensing and charge-coupling within the optical CCD sensor. During an appropriate exposure time the incident light generates electrons in each photosite, the number being proportional to the exposure time and the light intensity. A feedback process sets the exposure time. In low light levels this integration time is increased. In strong light conditions the amount of charge generated can be so large as to cause

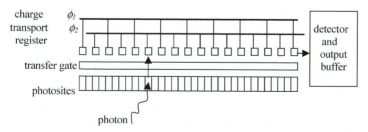

Figure 3.26. Schematic layout of an optical CCD sensor.

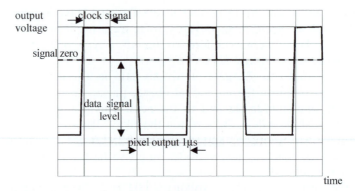

Figure 3.27. An optical CCD serial output signal.

leakage into adjacent photosites. When such saturation occurs the exposure time is automatically reduced.

After an appropriate exposure time the charge-packets in each of the photosites are all transferred in a single action to a CCD shift register mounted alongside the photosites. This transfer action is also accomplished by the charge-coupling process. A single, long electrode is used to provide the field switching-action to facilitate the transfer. This long gate-electrode is called the *transfer gate*. After a transfer has taken place the transfer gate is immediately returned to the non-transfer state so that the photosites can begin integrating the next image frame. During the light integration period the CCD transport register, now loaded with pixel data from the previous frame, is rapidly clocked in order to shunt the information to the device output circuitry. A serial set of analogue voltage pulses is output (figure 3.27).

The useful characteristics of optical CCDs are

- able to operate in low light levels,
- wide working temperature range of −55 to 70 °C,
- wide spectral response,
- low power consumption,
- low operating voltage,
- an output data rate of up to 20 MHz,
- automatic exposure system.

Devices are available from a number of companies in one- or two-dimensional format and with various numbers of cells and inter-cell spacings. For project work a small development board can be bought at low cost containing a one-dimensional array of 256 photosites, a cheap lens and additional circuitry. Currently, a mid-range two-dimensional device having 512×512 pixels can output 25 frames per second. Higher resolutions of 4096×4096 can only produce 0.5 frames per second. There are some very

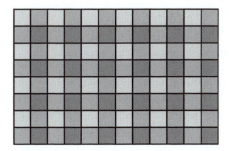

Figure 3.28. The Bayer filter pattern.

high frame-rate devices (1000 fps) available with correspondingly poor resolution (80 × 80 pixels). Colour imaging can be achieved in several ways.

Three-CCD system

The light beam is split using prisms so as to form three separate images. Each beam has either a red, green or blue filter in its path before reaching its particular CCD. Cameras using this approach produce rich, high resolution images of high quality but are bulky, expensive and have high energy consumption.

Light beam chopper

A single CCD chip is used. The image passes through a rotating wheel containing the red, green and blue filters. This is not a popular option since the colour information is produced sequentially and so high speed imaging is not possible.

Pixel filters

The individual photosites each have their own red, green or blue filter covering the entrance window. The filter layout is a mosaic pattern, the most common style being the Bayer pattern (figure 3.28).

Here photosites in alternate rows contain red and green filters and blue and green filters. The extra number of greens follows the spectral response of the eye. A so-called *de-mosaicing* algorithm is used on the data stream that renders an output colour tone for each pixel by averaging nearest neighbours. In this way resolution is not lost. This colour imaging system is the most popular in domestic quality cameras.

Variable depth photosites

A recent development employs the phenomenon that photons of different colour penetrate to different depths in silicon. Blue is the least penetrating and red the most. Photosites are stacked.

This variety of resolution quality and corresponding costs has meant that optical CCDs now enjoy a wide range of applications. These include

- facsimile machines using one-dimensional devices,
- high speed mail sorting,
- rapid non-contact inspection for quality control,
- computer-controlled material-handling systems,
- small, rugged, low power TV cameras for security use and remote inspection systems,
- digital photography,
- high performance video cameras.

3.6   Temperature sensors

The measurement of temperature is an important function in its own right. However, it is also possible to measure other variables by using heat flow. Such devices are termed *thermal* sensors. The primary sensor has its heat content reduced or increased by the variable and the secondary sensor measures the resulting temperature. An example of such a device is the vacuum pressure sensor. The cooling of a heated element by conduction to the enveloping residual gas causes a drop in its temperature. Element temperatures are calibrated against pressures for a specific gas. Other thermal sensors have been developed to measure flow, infrared radiation, gas analysis, true-rms power, humidity and dew point. Two types of infrared device are described below.

3.6.1   Thermal infrared radiation sensors

These devices operate on the warming effect of the incident radiation. This distinguishes them from devices such as infrared photodiodes where the quantum effects of individual photons is detected. However, they are orders of magnitude slower and less sensitive than the quantum devices but have a flat spectral response. They are usually used to remotely determine the temperature of incandescent objects. The energy absorbed causes a rise in the temperature of the sensor that can be related, using Planck's law, to the temperature of the hot object. Two common types of thermal detectors are described below.

Thermopiles and thermocouples

A thermopile is produced by mounting a number of thermocouples in series. All the hot junctions are tightly packed into the front or centre of the sensing head and the cold junctions to the cooler rear or rim. A black coating is the primary sensing element. Originally, bulk devices were not very precise and expensive. Miniature, semiconductor versions have been developed. Commercially available devices use a bismuth/antimony thermocouple (Elbel *et al* 1985), or a silicon/aluminium thermocouple (from Xensor Integration).

Silicon-based thermocouples are particularly attractive for inclusion in integrated circuit designs because standard integrated-circuit process technology can be used. Single-crystal, p-type silicon with aluminium generates approximately $1\,\text{mV}\,\text{K}^{-1}$. A distinguishing feature of the thermoelectric effect compared with other temperature measuring techniques is that a temperature difference is being measured. The cold junction will need to be carefully located on the chip. However, temperature difference is a variable that might be useful as part of another sensor type, for example gas flow as described in section 3.2.4.

Bolometers

These are devices that use a temperature-sensitive resistance. As above, a black coating is the primary sensor. The resistive material can be a thin film or flake of a metallic or semiconductor material, which is mounted on a non-conducting substrate such as a ceramic. The metals platinum, bismuth and telluride have been used. Values of order $100\,\Omega$ are typical. To achieve this, thin coatings of the material are evaporated on to the substrate in a pattern designed to give maximum track length. For high specification individual sensors are measured and trimmed to the correct tolerance. A glass envelope is applied over the surface for protection.

3.6.2 Thermistors

These are **therm**ally sensitive res**istors**. Devices first became available in the 1940s made from combinations of metal oxides. The proportions of the components are not a fixed ratio, as is the case with compounds, and consequently are referred to as non-stoichiometric substances. A negative temperature coefficient (NTC) is exhibited that can be extremely high $(-5\%\,\text{K}^{-1})$ compared with metals (see figure 3.29). The value of the coefficient, like the ambient temperature resistance, depends on the mixture of oxides used. Fabrication is by heating the compressed metal oxide powders at a sufficient temperature so that they sinter into a homogenous, polycrystalline solid. Oxides of manganese, nickel, copper, cobalt, vanadium, chromium, titanium and tungsten are used as well as various oxides of iron. Manufacturing limitations mean that the tolerances given to thermistor specifications are no better than a few percent. An alternative type of NTC device uses a single crystal of boron-doped silicon carbide. These have a lower sensitivity $(-2\%\,\text{K}^{-1})$ but better tolerances.

Generally, thermistors are not used for accurate thermometry because of their poor tolerance and high nonlinearity. They are used to make

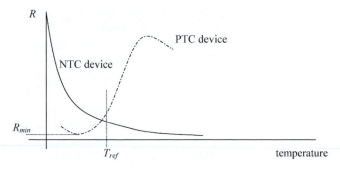

Figure 3.29. Thermistor characteristics.

thermometers of limited span and, because of their high sensitivity, for temperature control. The nonlinear relationship was first obtained by Steinhart and Hart in 1968 by curve-fitting. A polynomial of the form $1/T = A_0 + A_1 \ln R_T + \cdots + A_n(\ln R_T)^n$ was found to model the behaviour of a particular thermistor, where n is an integer, A_n are a set of constants and T is the absolute temperature. To a good approximation higher order terms can be ignored to give

$$R_T = K \exp\left(\frac{\beta}{T}\right) \tag{3.64}$$

where K and β are constants for the particular specimen. The form of equation (3.64) can also be obtained from solid-state physics using the Fermi–Dirac distribution on an ideal, uniform crystal.

For convenience equation (3.64) can be re-expressed to eliminate K using the thermistor's resistance at some ambient temperature R_a,

$$R_T = R_a \exp \beta\left(\frac{1}{T} - \frac{1}{T_a}\right). \tag{3.65}$$

PTC devices were introduced more recently following the discovery that the ceramic barium titanate when doped showed a distinct positive temperature coefficient. Other materials were developed using compounds of barium–strontium–lead titanate. Silicon PTCs are also available with somewhat better performance. The response curves are not as well behaved as the NTC devices so that an analytical solution can only be formulated for a restricted range. At some temperatures PTCs may operate as NTCs.

Generally, PTCs are used for temperature and over-current protection. Temperature protection is achieved by choosing a PTC whose R_{min} value occurs at a temperature just below the over-heat condition. It is mounted so as to be in good thermal contact with the system being monitored. For temperatures above T_{ref} the resistance increase can be as much as $100\% \,^{\circ}C^{-1}$. The characteristics published by manufacturers of PTCs often include the variation in resistance going through the values centred on T_{ref}.

An op-amp circuit can be constructed for buffering and enhancing the switching action. Alternatively, for low voltage applications the PTC can be placed directly into the power line of the system it is protecting. Consequently, heat transferred to the PTC from overheating of the system causes a severe current restriction. After the PTC has cooled normal operation resumes.

Over-current protection is achieved by placing the device in the power line of the system to be protected. Self-heating of the PTC occurs if the system demands excess current. Again the resulting increase in resistance causes the current to be restricted. Barium titanate is commonly used for this application in the form of a ceramic disc. Combinations of PTCs with resistors can be used to modify the characteristic.

For a more detailed description of thermistor characteristics and applications see Hyde (1971) and manufacturers' data sheets.

3.6.3   Integrated circuit temperature sensors

The electrical characteristic of the p–n diode junction is temperature sensitive. The ideal diode equation presented above in section 3.4.3 is reproduced below.

$$i_d = i_s \left[\exp \left(\frac{eV_d}{kT} \right) - 1 \right]. \tag{3.66}$$

Re-arranging gives

$$V_d = \frac{kT}{e} \ln \left(\frac{i_d}{i_s} + 1 \right) \tag{3.67}$$

which, in principle, indicates that if a constant current source is used to provide i_d then V_d is proportional to T. However, it is found from empirical data on the temperature sensitivity of the turn-on voltage (approximately 0.7 V at 25 °C for a silicon device) that the actual response differs from that predicted. From equation (3.67) the temperature sensitivity should go as

$$\frac{dV_d}{dT} = \frac{k}{e} \ln \left(\frac{i_d}{i_s} + 1 \right). \tag{3.68}$$

Taking the forward current as 10^3 larger than the reverse saturation current gives a theoretical temperature sensitivity of about $+0.6 \, \text{mV} \, \text{K}^{-1}$. The empirical result is a constant of $-2 \, \text{mV} \, \text{K}^{-1}$ over the range 50 K to 400 K. This anomaly is attributed to the temperature dependence of the reverse saturation current.

The base-emitter junction in the bipolar transistor can also be used in a similar way to the diode but tends to give better performance. So in practice the transistor, with its collector and base terminals joined, provides the better device. Generally, these p–n junction devices are referred to as PTATs, *proportional to absolute temperature.*

For the best PTAT performance two identical transistors are fabricated on the same chip and made to operate at a constant ratio r of their collector current densities. Under these conditions the difference in their base-emitter voltages is given by

$$\Delta V_{BE} = \left(\frac{kT}{e}\right) \ln r. \tag{3.69}$$

(See Sze 1994 for a typical circuit configuration). The name PTAT is normally associated with this type of construction.

A commercially available integrated circuit is the AD590 from Analog Devices. Low temperature-coefficient, thin-film resistors convert the PTAT voltage to a PTAT current. A linear output current of $1 \mu A K^{-1}$ is output over the range $-55\,°C$ to $150\,°C$. National Semiconductor also market an integrated-circuit temperature sensor, the LM135. It is described as 'operating as a 2-terminal Zener' with a breakdown voltage directly proportional to absolute temperature at $+10\,mV K^{-1}$.

Errors in PTAT devices can be as low as $0.25\,°C$ at room temperature and $0.75\,°C$ over the full operating range. They come in transistor package format, metal or plastic and 8-pin chip. This makes them well suited to circuit board mounting to provide temperature compensation or alarm. They can also be fastened to objects such as pipes and housings using thermally conducting cement. They have also been used in flow measurement, anemometry and liquid level measurement. Protection from moisture can easily be achieved using shrink-wrap sleeving. For more hostile environments sheathing or shrouds may be needed.

3.7   Detectors of ionizing radiation

3.7.1   Radioactivity and historical developments

Ionizing radiation occurs naturally from cosmic rays and whenever the structure of nuclei change. Disintegration, or *fission*, of a nucleus is termed radioactive decay and generally occurs with the heavier elements. Some unstable nuclei form part of a series. A series is a cascade of unstable, parent–daughter isotopes. There are four such series: three occur naturally and one is artificial. Three of the series end in stable isotopes of lead and the fourth in bismuth. During the decay process an α-particle or a β-particle is emitted and any excess energy is released as a γ-ray photon. Other particles which are much less ionizing may also be emitted, such as the neutron and the neutrino. Neutrons can be detected by secondary events such as nuclear scattering or transmutation. Changes to nuclei can also occur by the *fusion* of two lighter nuclei to form a heavier one. This process occurs in the stars, the hydrogen bomb and perhaps will be produced one day in a controlled way for electric power production. Particle accelerators and x-ray machines can also create ionizing radiation.

Ionizing radiation can be simply classified as one of two types: *photons*, which are part of the electromagnetic spectrum; and *particles*, such as the β-particle, a high energy electron (e^-), and the α-particle, a helium nucleus ($^4_2He^{2+}$).

The ionizing ability, and therefore the attenuation/absorption, is greatest with the heavy particles and least with the photons. Generally, this means that the detection and measurement of photons is the more difficult.

Radioactivity was discovered in Paris by Becquerel in 1896 and since then many techniques have been developed for the measurement of ionizing radiation. The discovery itself was made because samples of salts of uranium, being studied for their luminescent properties, exposed photographic plates stored in a lightproof container. Photographic film is still in use for x-ray imaging and dosimetry badges. The *ionization chamber* (1898) monitors current flow through an air or gas cell ionized by the radiation and can be calibrated to measure the energy of the particle or photon. The *Geiger–Muller tube* is a low-pressure variant of the ionization chamber. It is an event counter and is made more sensitive because of the avalanche produced. However, it gives no indication as to the energy of the particle or photon. The detection efficiency for photons is only 1%. The detected photons are absorbed by the wall material, which in turn produces secondary electrons that are detected in the usual way. In the *Wilson cloud* or *expansion chamber* (1911) super-saturated vapour condenses on air molecules ionized by the radiation path. The *hydrogen bubble chamber* (1952) contains super-heated liquid hydrogen in a pressurized tank having a glass top. Bubbles of hydrogen gas form around the ions produced by the radiation revealing the path. The original *scintillation counter* (c. 1900) was a screen coated with zinc sulphide (ZnS) used by Rutherford and others. The flashes of light emitted when struck by α-particles were counted manually in a darkened room over many hours. A modern, more convenient, device is the sodium iodide (NaI) detector activated with thallium. A photomultiplier is used to detect the light flashes produced by γ-ray photons. Caesiun iodide can also be used; it is less hygroscopic so can be used to detect α-particles by using an appropriate optically opaque cover. It has a higher absorption coefficient than NaI so thin layers can be used. It is also physically stronger, making it useful in space vehicles. A number of materials have been developed for scintillation use, including plastic sheets. A more recent development is the use of lithium fluoride (LiF) which, when absorbing high-energy photons, develops metastable states within the lattice. When heated later this energy is released as light (the thermoluminescent effect). An application is in re-useable dosimetry badges.

3.7.2 Counting and spectra

Generally, the higher the density of the material the more efficient it is at stopping the radiation. If the energy of an incident particle or photon is to be measured, rather than simply counting the number of events per second (the activity), then the radiation must be completely absorbed. Measuring the number of ion pairs produced by an event can then be used to calculate the initial energy of the particle or ray. Unless the activity is extremely high there will be no overlap of arrivals as the time within the detector is only 10^{-10} s. Such a procedure is called spectroscopy and the spectrum in figure 3.30 is typical of the time integration of arrivals. The broadening of the spectral lines is an artefact of the detection system.

3.7.3 Absorption processes

Charged particles

Kinetic energy is lost by collisions (which may produce ionization) or radiation. Energy loss is expressed as a function of path:

Loss of kinetic energy per unit path travelled

$$= -\mathrm{d}T/\mathrm{d}x \quad [\mathrm{J\,m^{-1}}] \tag{3.70}$$

$$= (-\mathrm{d}T/\mathrm{d}x)_{\mathrm{collision}} + (-\mathrm{d}T/\mathrm{d}x)_{\mathrm{radiation}}. \tag{3.71}$$

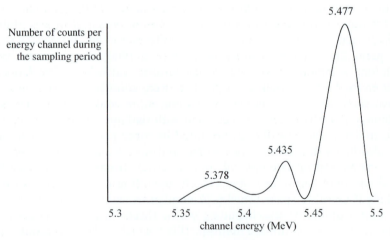

Figure 3.30. Energy spectrum of the α-particle source ^{241}Am.

The collision loss term is dependent on particle type, its instantaneous energy and the medium. Radiation loss is also energy dependent. The radiation type is usually Bremsstrahlung (breaking radiation) but for very high-energy particles Cherenkov radiation is also produced. This occurs when the speed of the particle is greater than the speed of light in the medium and is caused by the medium being polarized.

Photons

Numbers decline exponentially with path length. Intensity I, goes as

$$I = I_0 \exp(-\mu x) \tag{3.72}$$

where μ, the attenuation coefficient, is composed of

$$\mu = \mu_{es} + \mu_{pe} + \mu_{cs} + \mu_{pp} \tag{3.73}$$

where es is the *elastic scattering* out of the beam, pe is the *photoelectric effect* (the photon liberates an electron), cs is *Compton scattering* (the photon is scattered inelastically so that its energy decreases resulting in an increase in wavelength) and pp is *pair production* (the photon is annihilated to produce an electron and a positron).

Assuming that all the photon energy is lost within the detector, the elastic scattering can be ignored and μ can then be described as the absorption coefficient. Each of the absorption processes (photoelectric effect, Compton scattering and pair production) is energy and medium dependent.

3.7.4   Solid-state detectors

Particles, x-rays and γ-rays are commonly detected by PIN photodiodes. To produce an electron–hole pair in silicon requires an energy of about 3.6 eV. This is much lower than the typical 30 eV required to ionize a gas atom. Also, being a solid the high density means a higher probability of collision and hence detection. The concept of quantum efficiency η, as discussed in optical detection in section 3.4, again has to be considered. The efficiency depends on both the performance of the device in stopping the particle or photon and whether the electrons produced reach the electrodes. The fully reverse biased semiconductor diode has an excellent ability at separating and collecting the electron–hole pairs produced. Under these conditions η may be considered to be only dependent on the stopping ability of the device. Consequently, the structure may be modified in some applications to provide a larger sensing volume (figure 3.31).

High resistivity or compensated semiconductor material is essential, otherwise the number of free charge carriers already in the material will dominate those released by the radiation. Also, to ensure that all the charge

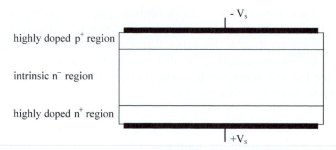

Figure 3.31. Photodiode structure for detecting radiation.

carriers produced reach the electrodes the material should be damage-free and contain minimal trapping centres.

Particle detection

Incident high-energy electrons of hundreds of keV produce typically 85 electron–hole pairs per μm in silicon. Each electron–hole pair created requires 3.6 eV, resulting in an energy loss of 300 eV μm^{-1}. Consequently, a silicon particle detector made 300 μm deep would produce about 4 fC of charge. For lower energy incident electrons detection is less efficient. This is because electrons generally suffer severe deflections at collisions, especially when colliding with heavy atoms. Thus, for lower incident speeds, the probability of backscatter out of the detector rises. Conversely, heavy ions are much more efficiently stopped so standard photodiodes can be used. These particles are stopped in the surface layers but this may mean that particles go undetected because of absorption in the entrance window. The detection of neutrons is not achieved directly but by secondary events following nuclear interactions. Neutron detectors may have a surface coating to generate secondary particles that are then detected by the diode.

Photon detection

High-energy photons interact with the detector material in one of the three ways described above in section 3.83. A typical response is shown in figure 3.32. For a more detailed description see Malm *et al* (1973).

For a photodiode in high reverse-bias as mentioned above all the charge carriers produced will be collected. Consequently, the quantum efficiency is set by the absorption efficiency so that equation (3.72) can be re-written using

$$\eta = \frac{I_0 - I}{I_0} \tag{3.74}$$

$$\eta = 1 - \exp(-\mu x). \tag{3.75}$$

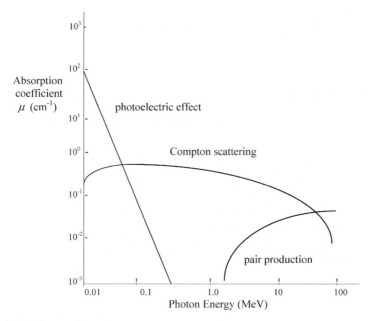

Figure 3.32. Photon absorption processes.

As an example the value for η, for a device 380 μm deep, on an n-type silicon substrate of resistivity 4000 Ω·cm, irradiated by 15 keV X-rays, is 55%.

Silicon and germanium devices have been in use since the 1960s and the technology for large, pure crystals has been well developed. Germanium has to be operated below 150 K to eliminate noise due to the effects of thermally produced leakage current. Research continues on material development and processing techniques. More recently cadmium telluride (CdTe) and mercury iodide (HgI_2) devices have become attractive. Their higher atomic number gives higher absorption coefficients but large crystals are difficult to produce without defects.

The energy resolution of photodiodes exceeds that of scintillation detectors by up to a factor of $\times 10$ and so produces more accurate energy spectra. However, scintillation detectors respond much faster and consequently can give lower counting errors.

CCDs for X-ray use

These are now being developed for satellite and medical use. Their high sensitivity makes them attractive, particularly in reducing patient radiation dose. Their small size means that they are gaining in popularity with orthodontists.

They are best used for low energy photons of order 10 keV. Their structure is similar to that described in section 3.5.4. JFETs are mounted on the CCD substrate so as to minimize the connection capacitance. Their high input impedance means there is good matching to the small output currents from the CCD. JFETs are also relatively immune to radiation damage.

3.8 Magnetic sensors

Magnetometers have been in use from the early days of electricity and magnetism. Early instruments were based on either the deflection of a compass needle, or the deflection of a ballistic galvanometer from the emf generated by a search coil, or more recently and still in common use, the flux-gate magnetometer. More exotic and sensitive devices have become available in recent years such as nuclear magnetic resonance magnetometers, capable of measuring nT, and the super-conducting interference device, or SQUID, used in biomagnetic research for measuring fields of pT.

For general laboratory use magnetometers are readily available for measurements of fields from 0.1 μT to several teslas. Table 3.3 indicates orders of magnitude for various sources of magnetic field.

Of the semiconductor based devices described below those based on the Hall effect are the most popular.

3.8.1 Hall effect

This phenomenon was discovered in 1879 by E H Hall. At the time conduction in metals was not understood (the electron itself was not discovered until 1897 by J J Thomson) and this effect proved to be very instructive.

Hall noted that a current-carrying conductor perpendicular to a magnetic field generates an emf in the mutually perpendicular direction, as shown in figure 3.33. The emf must be measured by a high-impedance voltmeter.

The explanation for the effect can be derived from the Lorentz force equation,

$$\boldsymbol{F} = q(\boldsymbol{E} + \boldsymbol{v} \wedge \boldsymbol{B}). \tag{3.76}$$

Table 3.3. Magnitudes of some common sources of magnetic fields.

Magnetic field source	Order of magnitude
Geomagnetism	10 μT
Data storage	1 mT
Current carrying conductor	1 mT
Small permanent magnet	10 mT
Superconducting magnet	10 T

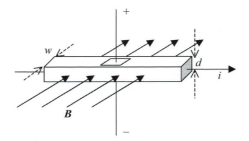

Figure 3.33. The Hall phenomenon.

In the derivation it is assumed that the length of the plate is much bigger than d. The magnetic field causes the moving charge carriers to move to the top of the plate of material. If the charges are not allowed to leave through the sides of the plate an equilibrium is reached where F is zero, giving

$$E_{\mathrm{H}} = -v \wedge B. \tag{3.77}$$

A dynamic equilibrium is reached where

$$\text{electrostatic repulsion} = \text{electromagnetic force}$$

E_{H} being the Hall electric field across the sides of the plate. The Hall voltage V_{H} can be obtained from

$$E_{\mathrm{H}} = \frac{V_{\mathrm{H}}}{d}. \tag{3.78}$$

Combining equations (3.77) and (3.78) gives

$$V_{\mathrm{H}} = -vBd. \tag{3.79}$$

Now, the current flowing in a material can be written in terms of the number density n of the charge carriers, the charge of a carrier q, the cross-sectional area A and the drift velocity v to give

$$i = nAqv. \tag{3.80}$$

Substituting for v in equation (3.79) and replacing A as wd gives

$$V_{\mathrm{H}} = -B\frac{i}{nAq}d \tag{3.81}$$

$$V_{\mathrm{H}} = -B\frac{i}{ndwq}d \tag{3.82}$$

$$V_{\mathrm{H}} = -B\frac{i}{nwq} \tag{3.83}$$

$$V_{\mathrm{H}} = R_{\mathrm{H}}B\frac{i}{w} \tag{3.84}$$

where R_{H} is called the Hall coefficient. In practice correction factors have to be added to equation (3.84) due to scattering of the charge carriers and geometrical effects. See Baltes and Castagnetti in Sze (1994) for a more detailed theoretical analysis.

At the time of the discovery it was found that most metals gave the expected result but some gave the opposite sign of V_H. Following the development of quantum mechanics and the band theory of conduction it was realized that these anomalous results were due to the charge carriers being holes.

In metals V_H is tiny and consequently has little practical use other than for materials research. However, the effect in semiconductors is orders of magnitude higher because of the lower value of n. Sensitivities can be between 5 and 500 V/AT, InSb being more sensitive than Si. Following the development of semiconductor fabrication techniques, Hall effect semiconductor probes began to be marketed as magnetometers.

There are a number of useful applications of the Hall effect:

- *Semiconductor research and development.* The Hall effect can be used to measure n and give the sign of the charge carriers to denote whether these are holes or electrons. Also, since the conductivity σ is determined by nqv, by measuring the conductivity v can be determined.
- *Magnetometer.* By making the connections as shown in figure 3.33 and driving a constant current through the plate the Hall voltage gives a direct reading of an unknown magnetic field. The side contact points need to be small and well aligned since a potential gradient exists along the plate, causing the current flow. Misalignment would generate a potential difference without a magnetic field being present. The magnetometer is small, compact, sensitive and can operate from d.c. up to 100 kHz. Currently, commercial products can be obtained with high sensitivity that measure down to 10 nT or, for high field strengths, alternative instruments can measure as high as 15 T. Various geometries are available. GaAs devices can operate at higher temperatures because intrinsic conduction only begins at about 400 °C in this III–V compound.
- *Proximity switch.* This is simply triggered by the presence of a magnetic field of sufficient strength. The plate of material has with it some TTL circuitry within the package. It is readily available from component suppliers. It is used for detecting metallic objects, for burglar alarm switches, and in some paddle wheel-type flowmeters.

3.8.2 Magnetoresistance

First discovered by W Thomson (later to become Lord Kelvin) in 1856, this phenomenon is also a result of the Lorentz force but it is the deflection of the current path by the magnetic field that is directly utilized here. In this device the length of the material is much shorter than its width so that the Hall field can be assumed to be negligible.

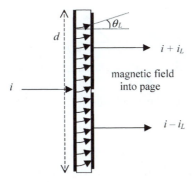

Figure 3.34. The magnetoresistive effect.

The magnetic field causes deflection of the current lines as shown in figure 3.34. This results in a current increase i_L through one side of the plate and a corresponding decrease in the other. The change in resistance of a specimen goes as

$$R = R_0(1 + \tan^2 \theta_L) \tag{3.85}$$

where θ_L is the Lorentz angle denoting the deflection of the current lines. It can be shown that $\tan \theta_L = \mu B$ where μ is the effective carrier mobility; see Baltes and Castagnetti in Sze (1994) for a fuller description.

A number of manufacturers produce magnetoresistive sensors. The usual fabrication is as a thin film of Permalloy (20% iron, 80% nickel) strips on a silicon substrate, which are connected so as to form a bridge network. This facilitates the production of bespoke configurations and integrated circuitry. Sensor range is typically ±5 mT with a resolution of 10 nT and developing a resistance change of about 3% at maximum field. Frequency response is similar to Hall effect devices at tens of kHz.

Materials exhibiting the so-called *giant magnetoresistive* or GMR *effect* discovered by Baibich *et al* (1988) are currently attracting a lot of attention. This effect is based on spin-dependent scattering of electrons within very thin sandwiched layers of materials. Alternate layers of Fe–Cr or Cu–Co are fabricated having thicknesses of about ten atoms. Resistance drops by up to 15% in fields of ±30 mT and bandwidth is high going from d.c. to MHz. Sensors based on GMR became available in the mid 1990s. Many magnetic storage media now employ GMR materials.

3.8.3 Magnetodiode

The magnetodiode structure is that of a long, thin PIN diode (figure 3.35). Double injection occurs from the n^+ and p^+ regions to give a high injection current into the intrinsic region. The carriers drift under the influence of the

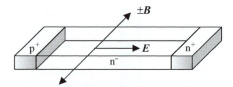

Figure 3.35. The magnetodiode PIN structure.

electric field **E** along the long axis. The presence of the magnetic field **B**, orthogonal to **E**, causes bunching of the charge carriers at the bottom or top faces of the plate depending on the direction of **B**. This is the so-called *magnetoconcentration* effect discovered by Suhl in 1949, which causes a gradient in the carrier concentration across the plate. The recombination rates at the two sides of the plate are arranged to be different. This has been achieved by polishing one side and rough grinding the other. The overall effect is to have the diode current–voltage characteristic modified by the magnetic field. Unfortunately this is a nonlinear relationship that is sensitive to temperature drift. Sensitivities of 5 V/mAT have been achieved at bias currents of around 5 mA and later versions, using CMOS technology, have produced 25 V/mAT. Frequency response is less than 10 kHz.

In production the quality control of the finishes to the plate surfaces is poor, meaning that carrier recombination rates vary so that reproducibility is poor. This together with the nonlinearity and temperature sensitivity has meant that there has been little commercial interest in magnetodiodes.

3.8.4 Magnetotransistor

These are transistors whose physical structure has been designed so that the collector current is optimally sensitivity to an applied magnetic field. A number of distinct phenomena are said to contribute to magnetotransistor operation. These are

- carrier deflection caused by the Lorentz force,
- magnetoconcentration,
- modulation of carrier injection.

The relative significance of these depends on the particular design.

An original device was the *drift-aided, lateral, double-collector* device developed by Davies and Wells (1970). The structure illustrated in figure 3.36 is of a bipolar p–n–p transistor.

The dominant process occurring here is carrier deflection. The base has current flowing through between the two end connections reminiscent of a Hall plate. The generated Hall field deflects the current carriers so that the two collectors receive differing amounts of current. The difference in the

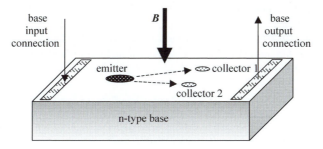

Figure 3.36. The drift-aided, lateral, double-collector, magnetotransistor.

collector currents Δi_c is given by

$$\Delta i_c = k(\mu_n + \mu_p)Bi_e \tag{3.86}$$

where k is a geometry factor, μ_n and μ_p are the electron and hole mobilities and i_e is the emitter current.

There have been many designs of magnetotransistors reported. Roumenin (1990) has given a classification and review of the various kinds. Some devices have the current flowing vertical to the surface unlike the lateral design shown above. These measure magnetic field in the plane of the structure.

Using equation (3.86) to define the sensitivity as $\Delta i_c/(Bi_e)$, a value of $0.25\,T^{-1}$ has been reported recently for a bipolar device employing both carrier deflection and injection modulation.

In addition to the standard bipolar construction technology, CMOS fabrication can be used. A typical sensitivity for a lateral CMOS device is $1.5\,T^{-1}$.

The MOSFET structure can be adapted so that the channel acts like a Hall plate or employs carrier deflection. These devices are termed MAGFETs. A recent report gave a sensitivity of 640 V/AT for a NMOS, Hall MAGFET operating at 5 V with a drain current of 0.5 mA.

Despite the advantage of higher sensitivity, magnetotransistors have not provided commercial competition to the inexpensive and simple Hall integrated circuit structure. For the various magnetotransistor designs this can be due to higher levels of noise, nonlinearity, temperature dependence and large off-set values that occur due to small manufacturing misalignment.

Bibliography

In this chapter various designs of novel semiconductor sensor have been described. However, there is little information on fabrication techniques. For those readers who would like to know more of this highly specialized foundry technology the following texts might be helpful.

Gardener J W, Varadan V K, Awadelkarim O O 2001 *Microsensors, MEMS and Smart Devices* (Wiley)
Kovacs G T A 1998 *Micromachined Transducers Sourcebook* (McGraw-Hill)
Maluf N 1999 *An Introduction to Microelectromechanical Systems Engineering* (Artech House)
Sze S M 1994 *Semiconductor Sensors* (Wiley)

References

Baibich R *et al* 1988 *Phys. Rev. Lett.* **61**(21) 2472–2475
Bernstein J *et al* 1993 'A micromachined comb-drive tuning fork rate gyroscope' *Proc. IEEE Micro Electromechanical Systems Workshop, Fort Lauderdale, FL*, February 1993, pp 143–148
Bouwstra S, Legtenberg R, Tilmans H A C and Elwenspeek M 1990 'Resonating micro-bridge mass flow sensor' *Sensors and Actuators* **A21–A23** 332–335
Cho S T and Wise K T 1993 'A high-performance microflowmeter with built-in self-test' *Sensors and Actuators A* **36** 47–56
Dalley J W and Riley W F 1978 *Experimental Stress Analysis* (New York: McGraw-Hill)
Davies L W and Wells M S 1970 'Magnetotransistor incorporated in a bipolar IC' *Proc. ICMCST, Sydney, Australia*, pp 34–43
Elbel T, Muller J E and Volklein F 1985 'Miniaturisierte thermische Strahlungsensoren: die neue Thermosaule TS-50.1' *Feingeratetechnik (Berlin)* **34** 113–115
Fairchild Weston CCD Sensors 1989 Systems and Development Technology, Databook
Gardner J W, Varadan V K and Awadelkarim O O 2001 *Microsensors, MEMS and Smart Devices* (Wiley)
Gardner J W 1994 *Microsensors: Principles and Applications* (Wiley)
Greenwood J C 1988 'Silicon in mechanical sensors' *J. Phys. E: Sci. Instrum.* **21** 1114–1128
Hanna S M 1987 'Magnetic field sensors based on SAW propagation in magnetic thin films' *IEEE Trans. UFFC* **34** 191–194
Johnson J D, Zarabadi S R and Sparks D R 1995 'Surface micromachined angular rate sensor' *Sensors and Actuators* **SP-1066** 77–83
Johnson R G and Higashi R E 1987 'A highly sensitive chip microtransducer for air flow and differential pressure sensing applications' *Sensors and Actuators* **11** 63–72
Kovacs G T A 1998 *Micromachined Transducers Sourcebook* (McGraw-Hill)
Kress H-J, Marek J, Mast M, Schatz O and Muchow J 1995 'Integrated silicon pressure sensor for automotive application with electronic trimming' *Sensors and Actuators* pp 35–40
Malm H L, Randorf T W, Martini M and Zanio K R 1973 Gamma ray efficiency comparisons for Si(Li), Ge, CdTe and HgI$_2$ detectors *IEEE Trans. Nucl. Sci.* **20** 500–509
Maluf N 1999 *An Introduction to Microelectromechanical Systems Engineering* (Boston: Artech House)
Neubert H K P 1975 *Instrumentation Transducers: An Introduction to their Performance and Design* 2nd edition (London: Oxford University Press) pp 237–238
Neumeister J, Thum R and Lüder E 1990 'A SAW delay-line oscillator as a high-resolution temperature sensor' *Sensors and Actuators A* **21–23** 670–672
Patel H 1992 'Silicon that moves' *IEE Review* July/August pp 268, 269
Reidemeister E P and Johnson L K 1995 'Capacitive acceleration sensor for vehicle applications' *Sensors and Actuators* **SP-1066** pp 29–34

Roumenin C S 1990 'Bipolar magnetotransistor sensors. An invited review' *Sensors and Actuators* **A24** 83

Spark D R, Zarabadi S R, Johnson J D, Jiang Q, Chia M, Larsen Q, Higdon W and Castillo-Borelley P 1997 'A CMOS integrated surface micromachined angular rate sensor: its automotive applications' *Proc. Transducers '97, Int. Conf. on Solid State Sensors and Actuators, Chicago, IL*, 16–19 June, pp 851–854

Sze S M 1981 *Physics of Semiconductor Devices* 2nd edition (Wiley)

Sze S M 1994 *Semiconductor Sensors* (Wiley)

Wilson J and Hawkes J F B 1983 *Optoelectrics: An Introduction* (Prentice-Hall)

Xensor Integration, Delft, the Netherlands, Data sheet XI-IR3774 Infrared sensor, 91.083774

Exercises

1. Explain why semiconductor-based sensors are at the centre of many research and development programmes. Describe some of the difficulties in applying such semiconductor devices to the various measurement fields.

2. Discuss the advantages and disadvantages to using silicon as a construction material for micro sensors.

3. A miniature gas flow sensor is constructed using a silicon nitride bridge-structure operating as a resonant system.
   (i) Describe the operating principle.
   (ii) A piezoresistive strip is attached to the structure to monitor the oscillations. Draw a functional block diagram to illustrate how resonance might be maintained.

4. Pressure sensors are now available manufactured from a silicon disc. The disc acts as a diaphragm on which piezoresistive tracks have been formed.
   (i) State two restrictions that limit the applications of such devices.
   (ii) The radial and circular stresses occurring in the disc are given by

$$\text{radial stress} \qquad \varepsilon_r = \frac{3P(1-v^2)}{8Et^2}(R_0^2 - 3r^2)$$

$$\text{tangential stress} \qquad \varepsilon_t = \frac{3P(1-v^2)}{8Et^2}(R_0^2 - r^2).$$

   For a fixed pressure, sketch the forms of these stress relationships as a function of the radius. Where on the diaphragm is the point of zero stress?
   (iii) Draw an appropriate signal conditioning circuit and explain how the piezoresistive elements need to be connected to provide temperature compensation and give maximum sensitivity.

5. (a) Explain the phenomenon of the surface acoustic wave giving typical numerical values of the associated parameters. Describe how a surface acoustic wave sensor might be constructed to measure
   (i) pressure
   (ii) magnetic field.
   (b) A certain surface acoustic wave mass sensor has a sensitivity given by

$$\frac{df}{d\mu} = (k_1 + k_2)f_0^2 \quad [\text{Hz kg}^{-1}\,\text{m}^2]$$

Given that the resonant frequency is 10^9 Hz and that

$$k_1 = -9.8 \times 10^{-8} \quad \text{and} \quad k_2 = -4.18 \times 10^{-8}\,\text{m}^2\,\text{s kg}^{-1}$$

determine the resolution of the mass measurement if the signal processing system can measure to a resolution of 5 in 10^8.

6. Explain the terms in the modified Nernst equation

$$E_{\text{Thévenin}} = E_0 + \frac{RT}{zF}\ln(a_x + ka_y).$$

Given that

$$\text{pH} = -\log_{10} a_\text{H}$$

combine these two relationships to show that

$$E_{\text{Thévenin}} = E_0 - 0.0592\,\text{pH} \quad [\text{volts}]$$

7. Describe, with the aid of a diagram, the operating principles of a FET adapted for use as a biosensor. Give some advantages and disadvantages of the invasive use of these sensors with patients.

8. (a) Draw a circuit diagram for a photodiode and its signal conditioning so that it functions in the photovoltaic mode. Show that in this mode

$$V_\text{d} = \frac{kT}{e}\ln\left(\frac{e\eta PA\lambda}{i_s hc}\right).$$

What are the advantages and disadvantages to this mode of operation?
   (b) (i) Photodiodes can also be used to detect x-rays. The absorption of the rays can be described by

$$I = I_0 \exp(-\mu x).$$

Show that

$$\eta = 1 - \exp(-\mu x).$$

   (ii) A certain photodiode used to detect x-rays has a sensor depth of 300 µm and the value of μ is $2 \times 10^3\,\text{m}^{-1}$. Calculate the photocurrent given that 85 electron–hole pairs are produced per µm and that the incident photon flux is $10^5\,\text{s}^{-1}$.

9. (a) Describe the fabrication and performance of the NTC and PTC thermistors. Include sketch graphs and applications.

   (b) Using the ideal diode equation show that the forward bias voltage when a diode is conducting responds PTAT.

   (c) The p–n junction in a PTAT device is usually a transistor base-emitter junction rather than a diode. Why is this?

   (d) One design of a PTAT device uses two identical transistors mounted on the same substrate. Their collector currents i_{c1} and i_{c2} are arranged to be in a fixed ratio γ. Using your expression from (b) above show that the difference in their base currents ΔV_{be} goes as

$$\Delta V_{be} = \frac{kT}{e} \ln \gamma.$$

10. A certain magnetoresistor is commercially available having a resistance R_0 of $500.0\,\Omega$ and a sensitivity of $2.5\,k\Omega\,T^{-1}$.

    (i) Draw the physical layout of a bridge circuit similar to that of figure 3.4 such that four of these magnetoresistors might be used to measure a magnetic field.

    (ii) Show that for small changes in resistance δR and a supply voltage V_s, the optimum Thévenin output potential difference is given by

$$E_{\text{Thévenin}} = \frac{\delta R}{R_0} V_s.$$

    (iii) Assuming that noise levels in the circuitry might be in the region of $50\,\mu V$ estimate the minimum magnetic field that could be measured.

Chapter 4

Optical fibre sensors

4.1   Introduction

The first patents for optical fibre sensors appeared in the mid-1960s. Since then significant development has taken place from these first simple, light-guide devices to the present intricate and sophisticated systems. The advance has been facilitated by the rapid progress made in the use of optical fibres by the telecommunications industry. This has resulted in the availability of high-quality, cheap fibres. In addition, the growth in the use of compact disc players, laser printers and infrared remote control units has made optical components readily available at low cost.

The establishment of optical fibres for telecommunication links has encouraged their use for telemetry in some industrial fields. The chemical, nuclear, petrochemical and other process industries have begun to use them for supervisory control and data acquisition (SCADA) principally because of their

- lower installation costs,
- intrinsic safety due to the electrical isolation and low-power light levels,
- robustness and reliability, leading to lower maintenance costs,
- immunity to electromagnetic interference,
- chemical inertness,
- absence of electromagnetic transmission,
- high temperature resilience,
- high bandwidth.

Industries such as electricity, automobile and avionics have also shown interest, as have robotic and medical engineers.

The use of optical fibres in instrumentation systems can be grouped into three categories,

- light guides for illumination in remote monitoring and in obscure locations,

112

- imaging systems, such as endoscopes, face-plates and magnifiers,
- sensors that use the fibre itself or some small artefact coupled to the fibre.

It is this last category that is attracting most effort currently and is the theme of this chapter. The concept of combining an optical fibre sensor and the data link as a complete system is very attractive, especially to industrial instrumentation engineers. The advantages listed above for the optical fibre data link should then become true for the whole instrumentation system. In addition, the low weight and the small size of optical fibre sensors are attractive features in many measurement fields.

The common parameters to be measured in the process industries are temperature, flow, pressure, displacement/position and levels in tanks and hoppers. It is unlikely that sophisticated and expensive devices will be taken up initially. The industry is traditionally very conservative and needs much convincing. The harsh environment of process plant will make delicate devices impractical. It is most likely that the technology will gain acceptance through the introduction of simple sensor systems that are relatively cheap and reliable or highly specified and unique. For multi-sensing situations a simple radial network of links to the control room will be used, although a data highway ring is envisaged for future, more extensive systems.

4.2   The optical waveguide

Telemetry based on metallic wire pairs, although by far the most common technique currently, is facing a growing challenge from optical fibres in all applications. Light is, without doubt, one of the most important physical quantities in the universe. The Biblical creation story records the creation of light as one of the earliest acts of God. There is also little doubt that light was the original historic method for transmitting information over a distance. This would have been achieved by reflection of sunlight, lamps or fire beacons. At the end of the 6th century BC Aeschylus recorded that the news of the downfall of Troy was relayed from Asia Minor to Argos using a chain of fire signals. In the 2nd century BC Polybius described a method by which the Greek alphabet could be transmitted with fire signals using a two-digit, five-level code. In 1870 John Tyndall demonstrated that light could follow a curved jet of water and later J L Baird filed patents for the transmission of light in glass rods. In the 1950s fibre optic devices were developed for inspection instruments. However, it was in 1966 that Charles Kao and George Hockham at Standard Telecommunication Laboratories, Harlow, proposed the principle of information transmission via a transparent dielectric. It was the subsequent development of high quality glass fibres that led to their current widespread use in telecommunications.

Recent developments in data compression techniques have made possible the transmission of real-time, digital, video images in serial format. This advance, with the intrinsic benefits of optical fibres, has accelerated the phenomenal growth of an international, information transmission industry that struggles to keep up with the increasing demand for capacity. A recent development, as a means of increasing capacity without installing extra lines, is the use of light of different wavelengths down a fibre to provide extra channels, in effect frequency domain multiplexing. It is usually referred to as *dense wavelength division multiplexing* or DWDM. The current maximum in commercial use is 100 different wavelengths per fibre. With a transmission rate of 10 gigabits s^{-1} per channel this yields a total single-fibre rate of 1 terabit (10^{12}) s^{-1}. Gupta *et al* (2002) have reported a laboratory rate of 3.2 terabits s^{-1} using 40 lasers in the range 1531–1563 nm. The channel bit-rate was 80 gigabits s^{-1} over 100 km.

The cell network used by cell phones employs optical fibres for interlinking. Mobile cell phones themselves are only short range and register with a GHz antenna system in the local cell. The use of mobile phones for games and video imaging is further fuelling demand for capacity. Clearly, the level of utilization of optical fibre telecommunications is set to increase well into the decades ahead.

4.2.1   Principles of propagation

An optical fibre is a waveguide usually of circular cross-section. The fibre is made from some dielectric material such as a polymer or a glass. To understand the propagation process requires an appreciation of classical electromagnetic wave theory. It was James Clerk Maxwell in 1864 who showed conclusively that light was an electromagnetic phenomenon. This was an idea originally proposed by Michael Faraday. Maxwell was able to show that by combining the known magnetic and electrostatic field equations, with an important modification, a pair of differential equations could be produced whose solutions were transverse waves. Moreover, the velocity of these waves was the same as the free-space speed of light. This confirmed that what is now called the electromagnetic spectrum is indeed composed of electric and magnetic fields. This derivation of Maxwell's is found in many texts and is illustrated below.

Gauss's theorem

In differential equation format, Gauss's electric field equation for free space can be expressed in cartesian coordinates as

$$\frac{\partial E_x}{\partial x} + \frac{\partial E_y}{\partial y} + \frac{\partial E_z}{\partial z} = \frac{\rho}{\varepsilon_0} \tag{4.1}$$

where ρ is the free charge density. For convenience this is often expressed more generally in terms of a vector differential operator ∇, pronounced *del*. Del can be interpreted in any three-dimensional co-ordinate system. In the cartesian system it is expressed as

$$\nabla \equiv \left(i\frac{\partial}{\partial x} + j\frac{\partial}{\partial y} + k\frac{\partial}{\partial z}\right) \tag{4.2}$$

where i, j and k are unit vectors in the x, y and z directions respectively. Del can operate on a scalar or a vector function. The non-vector operation on the scalar ϕ is

$$\nabla\phi = \left(i\frac{\partial}{\partial x} + j\frac{\partial}{\partial y} + k\frac{\partial}{\partial z}\right)\phi. \tag{4.3}$$

This scalar operation is the gradient of ϕ and is sometimes written *grad ϕ*. When operating on a vector function the multiplication can be one of two forms. These are the *dot* product $\cdot$, or *cross* product $\times$. The dot product on some vector Y in cartesian coordinates is written

$$\nabla \cdot Y = \left(i\frac{\partial}{\partial x} + j\frac{\partial}{\partial y} + k\frac{\partial}{\partial z}\right) \cdot Y. \tag{4.4}$$

This is sometimes referred to as the divergence of Y and written *div Y*. Dot product multiplication produces a scalar. The alternative cross product vector multiplication is

$$\nabla \times Y = \left(i\frac{\partial}{\partial x} + j\frac{\partial}{\partial y} + k\frac{\partial}{\partial z}\right) \times Y. \tag{4.5}$$

$\nabla \times Y$ is also-called curl Y. The resulting quantity is a vector normal to the plane formed by the original vectors.

Following this notation equation (4.1) can be written

$$\text{div } E = \frac{\rho}{\varepsilon_0}. \tag{4.6}$$

If a dielectric medium is present then the presence of induced charge has to be included in the calculation of E. The introduction of the vector D, called the *electric displacement*, overcomes this problem. D is only dependent on the free charges present and is defined as

$$D = \varepsilon_r\varepsilon_0 E \quad \text{or} \quad \varepsilon E. \tag{4.7}$$

Thus equation (4.6) becomes

$$\text{div } D = \rho. \tag{4.8}$$

If there are no free charges present, as is the case for an optical fibre, then

$$\text{div } D = 0. \tag{4.9}$$

A similar expression can be written for the magnetic field B as

$$\text{div } B = 0. \tag{4.10}$$

Div B is always zero because there are no free magnetic poles. The magnetic field B is related to the magnetization force H by

$$B = \mu_0 \mu_r H \quad \text{or} \quad \mu H. \tag{4.11}$$

H unlike B is not altered by the presence of some material. This is similar to the vector D, which is only dependent on the presence of real charges and not on any induced charge in the medium.

Faraday's law

Faraday's law of electromagnetic induction can be expressed in this vector calculus form as

$$\nabla \times E = -\frac{\partial B}{\partial t} \tag{4.12}$$

the negative sign coming from Lentz's law. Using the description of curl given after equation (4.5), equation (4.12) can be written

$$\text{curl } E = -\frac{\partial B}{\partial t}. \tag{4.13}$$

Ampère's theorem

Ampère's theorem relates the magnetization force H to the actual current flowing I, taken around some closed path s,

$$\oint H \cdot \mathrm{d}s = I. \tag{4.14}$$

This can be expressed in differential equation form in terms of the free current density J as

$$\text{curl } H = J. \tag{4.15}$$

However, as mentioned above, Maxwell introduced an extra term to incorporate any current due to the electric displacement D. Equation (4.15) becomes

$$\text{curl } H = \frac{\partial D}{\partial t} + J. \tag{4.16}$$

Because there are no free charges in an optical fibre equation (4.16) reduces to

$$\text{curl } H = \frac{\partial D}{\partial t}. \tag{4.17}$$

Maxwell's equations

The four equations that have become know as Maxwell's equations are obtained using equations (4.7) and (4.11) to substitute for D and B so that equation (4.9) becomes

$$\text{div } \boldsymbol{E} = 0 \tag{4.18}$$

equation (4.10) becomes

$$\text{div } \boldsymbol{H} = 0 \tag{4.19}$$

equation (4.13) becomes

$$\text{curl } \boldsymbol{E} = -\mu \frac{\partial \boldsymbol{H}}{\partial t} \tag{4.20}$$

and equation (4.17) becomes

$$\text{curl } \boldsymbol{H} = \varepsilon \frac{\partial \boldsymbol{E}}{\partial t}. \tag{4.21}$$

The final stage in the derivation involves combining these four equations to solve for $\boldsymbol{E}$ and $\boldsymbol{H}$. To do this the following mathematical identity is required:

$$\nabla \times (\nabla \times \boldsymbol{Y}) = \nabla(\nabla \cdot \boldsymbol{Y}) - \nabla^2(\boldsymbol{Y}). \tag{4.22}$$

∇^2 is the second derivative of ∇ and is sometimes referred to as the Laplacian operator. Thus, taking the curl of curl $\boldsymbol{E}$ gives

$$\text{curl curl } \boldsymbol{E} = \text{grad div } \boldsymbol{E} - \nabla^2 \boldsymbol{E} \tag{4.23}$$

$$\text{curl}\left(-\mu \frac{\partial \boldsymbol{H}}{\partial t}\right) = \text{grad } 0 - \nabla^2 \boldsymbol{E} \tag{4.24}$$

$$\mu \frac{\partial(\text{curl } \boldsymbol{H})}{\partial t} = \nabla^2 \boldsymbol{E} \tag{4.25}$$

$$\mu\varepsilon \frac{\partial(\partial \boldsymbol{E}/\partial t)}{\partial t} = \nabla^2 \boldsymbol{E} \tag{4.26}$$

$$\nabla^2 \boldsymbol{E} = \mu\varepsilon \frac{\partial^2 \boldsymbol{E}}{\partial t^2}. \tag{4.27}$$

Now the general expression for a wave of amplitude $\boldsymbol{Y}$ in differential equation format is

$$\nabla^2 \boldsymbol{Y} = \frac{1}{v_{\text{p}}^2} \frac{\partial^2 \boldsymbol{Y}}{\partial t^2} \tag{4.28}$$

where v_{p} is the phase velocity of the wave, the velocity at which the wave crests move forward. By comparing terms it can be seen that equation (4.27) is a wave and that

$$v_{\text{p}} = \frac{1}{\sqrt{\mu\varepsilon}}. \tag{4.29}$$

For free space equation (4.29) becomes

$$c = \frac{1}{\sqrt{\mu_0\varepsilon_0}}. \tag{4.30}$$

Inserting the values of μ_0 and ε_0 in equation (4.30) gives the free-space speed of light, which added confirmation to Maxwell's theory.

If the derivation is carried out for $\boldsymbol{H}$ a similar expression to equation (4.27) is obtained,

$$\nabla^2 \boldsymbol{H} = \mu\varepsilon \frac{\partial^2 \boldsymbol{H}}{\partial t^2}. \tag{4.31}$$

The general solution to equation (4.28) is the sinusoidal wave equation

$$\boldsymbol{Y} = \boldsymbol{Y}_0 \exp j(\omega t - \boldsymbol{k} \cdot \boldsymbol{s}) \tag{4.32}$$

where $\boldsymbol{k}$ is the propagation vector and $\boldsymbol{s}$ is the displacement. The magnitude of $\boldsymbol{k}$ is given by

$$|\boldsymbol{k}| = k = \frac{2\pi}{\lambda_0}. \tag{4.33}$$

k is also-called the free-space wave number. From equation (4.32) the sinusoidal wave equations for $\boldsymbol{E}$ and $\boldsymbol{H}$ are

$$\boldsymbol{E} = \boldsymbol{E}_0 \exp j(\omega t - \boldsymbol{k} \cdot \boldsymbol{s}) \tag{4.34}$$

$$\boldsymbol{H} = \boldsymbol{H}_0 \exp j(\omega t - \boldsymbol{k} \cdot \boldsymbol{s}). \tag{4.35}$$

Orientation of the fields in an electromagnetic wave

Inserting equations (4.34) and (4.35) into the four Maxwell equations yields

$$\boldsymbol{k} \cdot \boldsymbol{E} = 0 \tag{4.36}$$

$$\boldsymbol{k} \cdot \boldsymbol{H} = 0 \tag{4.37}$$

$$\boldsymbol{k} \times \boldsymbol{E} = \mu\omega \boldsymbol{H} \tag{4.38}$$

$$\boldsymbol{k} \times \boldsymbol{H} = -\varepsilon\omega \boldsymbol{E}. \tag{4.39}$$

Equations (4.36) and (4.37) reveal that the electric and magnetic fields are orthogonal to the direction of travel. Equation (4.38) shows that $\boldsymbol{H}$ is normal to the plane of $\boldsymbol{k}$ and $\boldsymbol{E}$ and equation (4.39) that $\boldsymbol{E}$ is normal to the plane of $\boldsymbol{k}$ and $\boldsymbol{H}$. For these conditions all to be true $\boldsymbol{k}$, $\boldsymbol{H}$ and $\boldsymbol{E}$ must be mutually orthogonal as shown in figure 4.1.

The laws of reflection and refraction

The equations derived above can be used to analyse the phenomena that occur as a wave crosses the boundary between two different dielectric materials. This is a significant issue with optical fibres since the fibre must have a dielectric outer cladding to support and protect it. Figure 4.2 illustrates a wave travelling with velocity v_1 in the first medium and v_2 in the second.

The three waves in the two media have the propagation vectors $\boldsymbol{k}_i$ for the incident wave, $\boldsymbol{k}_{re}$ for the reflected wave and $\boldsymbol{k}_{ra}$ for the refracted wave. The

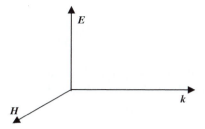

Figure 4.1. The mutual orthogonality of the electric and magnetic fields and the direction of propagation.

wave equations are

$$Y = Y_0 \exp j(\omega t - k_i \cdot s) \tag{4.40}$$

$$Y = Y_0 \exp j(\omega t - k_{re} \cdot s) \tag{4.41}$$

$$Y = Y_0 \exp j(\omega t - k_{ra} \cdot s) \tag{4.42}$$

where Y represents E or H.

In the plane of the boundary all three components of the waves must be in phase otherwise there would be a discontinuity. This is true for any wave phenomenon. Thus,

$$k_i \cdot s = k_{re} \cdot s = k_{ra} \cdot s \tag{4.43}$$

which becomes

$$k_i \sin \alpha = k_{re} \sin \beta = k_{ra} \sin \gamma \tag{4.44}$$

The magnitudes of k_i and k_{re} are the same since they are in the same material and so will have the same wavelength λ. Consequently,

$$\angle \alpha = \angle \beta \tag{4.45}$$

one of the laws of reflection. Also, writing equation (4.33) as

$$|k| = k = \frac{2\pi}{\lambda_0} = \frac{2\pi f}{c} \tag{4.46}$$

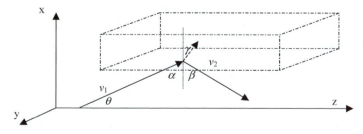

Figure 4.2. Transmission of a wave across a dielectric boundary.

equation (4.44) can be re-written as

$$\frac{2\pi f}{v_1}\sin\alpha = \frac{2\pi f}{v_2}\sin\gamma. \tag{4.47}$$

The frequency does not alter across the boundary otherwise a discontinuity would occur as stated above. Consequently, equation (4.47) becomes

$$\frac{\sin\alpha}{\sin\gamma} = \frac{v_1}{v_2} = {}_1 n_2 \tag{4.48}$$

which is Snell's law, ${}_1n_2$ being the refractive index going from medium 1 to medium 2. If medium 1 is free space then equation (4.48) becomes

$$\frac{\sin\alpha}{\sin\gamma} = \frac{c}{v_2} = n_2 \tag{4.49}$$

where n_2 is the absolute refractive index of medium 2. It is often more convenient to deal with absolute refractive indices. Equation (4.48) can be re-written

$$\frac{\sin\alpha}{\sin\gamma} = \frac{c/v_2}{c/v_1} = \frac{n_2}{n_1}. \tag{4.50}$$

If the wave is to travel down the optical fibre without loss then the angle of incidence at the walls α must be large enough for total internal reflection to occur. The least angle of incidence for this to occur is when

$$\sin\gamma = 1 \tag{4.51}$$

$$\gamma = \frac{\pi}{2}. \tag{4.52}$$

From equation (4.50), it can be seen that this particular value of α, called the *critical angle*, is $\sin^{-1} n_2/n_1$.

The requirement to have waves strike the inside of the core fibre at greater than the critical angle has implications on the entrance and exit of light at the fibre ends. This is indicated in figure 4.3.

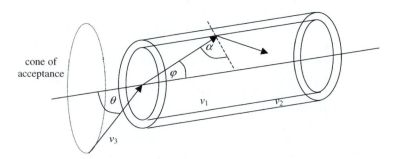

Figure 4.3. The numerical aperture.

At the entrance to the fibre, Snell's law, equation (4.48), becomes

$$\frac{\sin \theta}{\sin \varphi} = \frac{n_1}{n_3} \qquad (4.53)$$

where n_3 is the refractive index in the medium adjacent to the fibre end. Re-arranging equation (4.53) gives

$$\sin \theta = \frac{n_1}{n_3} \sin \varphi. \qquad (4.54)$$

But from figure 4.3 it can be seen that $\sin \varphi = \cos \alpha$, so equation (4.54) becomes

$$\sin \theta = \frac{n_1}{n_3} \cos \alpha. \qquad (4.55)$$

If α is the critical angle then $\sin \alpha = n_2/n_1$. Using the trigonometrical relationship $\cos^2 \alpha = 1 - \sin^2 \alpha$ equation (4.55) becomes

$$\sin \theta = \frac{n_1}{n_3} \sqrt{1 - \sin^2 \alpha} \qquad (4.56)$$

$$\sin \theta = \frac{n_1}{n_3} \sqrt{1 - \left(\frac{n_2}{n_1}\right)^2} \qquad (4.57)$$

$$\sin \theta = \frac{1}{n_3} \sqrt{n_1^2 - n_2^2} \qquad (4.58)$$

$$n_3 \sin \theta = \sqrt{n_1^2 - n_2^2}. \qquad (4.59)$$

The expression $n_3 \sin \theta$ is called the numerical aperture, NA.

$$\mathrm{NA} = \sqrt{n_1^2 - n_2^2}. \qquad (4.60)$$

NA becomes smaller as the value of n_2 approaches that of n_1. In commercial fibres the difference δ between n_1 and n_2 is only about 1%, making NA about 0.2. Increasing δ has the disadvantage of increasing inter-modal dispersion, a phenomenon described later.

For light to propagate along the fibre it must enter within a cone of half-angle θ as shown in figure 4.3. This is sometimes referred to as the *cone of acceptance*. Also, because in geometrical ray diagrams rays are reversible, an identical cone must exist at the emerging end of the fibre within which the light is confined. These two cones indicate the optimum position relative to the ends of the fibre for the light source and detector (figure 4.4). In addition, the emitting area of the light source should be less than the fibre cross-sectional area and be close to the fibre end. Also, the emission half-angle of the source should be less than that of the fibre. The sensitive surface of the detector should be large enough and placed such that all of the emergent light is captured.

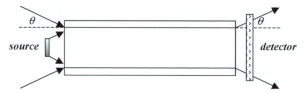

Figure 4.4. Placement of light source and detector.

The concepts described above of reflection, refraction, critical angle and cones of acceptance and emergence, apply to fibres of core diameter $>10\,\mu m$. The transmission paths through such fibres are numerous, each suffering a different number of reflections. Such fibres are classed as *multimode*. For smaller fibres, whose diameters are approaching the wavelength of the light being transmitted, interference effects begin to occur. This has the result of reducing the number of modes of transmission. In the smallest of fibres used only one mode of transmission is possible. The wave travels along the central axis of the fibre and is called the *meridional* ray. These *monomode* fibres are preferred in long distance telemetry because a light pulse will be elongated in time if the light travels along multiple paths of various lengths. Pulse broadening, or *dispersion*, sets an upper limit on the bit-rate that can be serially transmitted. There are other causes of dispersion so this form is called intermodal or mode dispersion. Graded refractive index optical fibre is an alternative to monomode. It is of a similar diameter to multimode but light travelling near the edges of the rod travels faster than the axial rays and so compensation is achieved. It is the less commonly used type.

Graded refractive index (GRINrod) lenses have been developed for optical fibre systems and are often used in sensor research. These are rod-shaped, similar to the fibres, but are usually between 1 and 5 mm in diameter whereas multimode and graded index fibres can be up to $100\,\mu m$ in diameter although $50\,\mu m$ is the more typical.

The evanescent wave

Although there is no refracted wave once the critical angle has been exceeded this does not mean that there is no disturbance in the cladding medium. Considering again the refracted wave of figure 4.2 given by equation (4.42) and assuming it to be plane polarized in the x direction, resolving into the x and z directions gives

$$Y = Y_0 \exp j(\omega t - (k_{\mathrm{ra}} x \cos \gamma + k_{\mathrm{ra}} z \sin \gamma)). \tag{4.61}$$

Now, re-arranging equation (4.50) gives an expression for $\sin \gamma$,

$$\sin \gamma = \frac{n_1}{n_2} \sin \alpha. \tag{4.62}$$

But equation (4.62) implies that, for angles of α greater than the critical angle, $\sin \gamma > 1$, which is not possible. To continue this analysis it is therefore expedient to replace both $\sin \gamma$ and $\cos \gamma$ in equation (4.61):

$$Y = Y_0 \exp j \left(\omega t - \left(k_{\text{ra}} x \sqrt{1 - \sin^2 \gamma} + k_{\text{ra}} z \frac{n_1}{n_2} \sin \alpha \right) \right). \tag{4.63}$$

But since $\sin \gamma > 1$, the root here is imaginary so the complex form is substituted:

$$\sqrt{1 - \sin^2 \gamma} = \pm j \sqrt{\sin^2 \gamma - 1}. \tag{4.64}$$

Equation (4.61) becomes

$$Y = Y_0 \exp j \left(\omega t - \left(\pm j k_{\text{ra}} x (\sin^2 \gamma - 1)^{1/2} + k_{\text{ra}} z \frac{n_1}{n_2} \sin \alpha \right) \right). \tag{4.65}$$

The physically more meaningful solution is obtained by taking the negative root to give an exponential decline with x,

$$Y = Y_0 \exp(-k_{\text{ra}} x (\sin^2 \gamma - 1)^{1/2}) \exp j \left(\omega t - k_{\text{ra}} z \frac{n_1}{n_2} \sin \alpha \right). \tag{4.66}$$

This can be expressed as

$$Y = Y_0 \exp(-x/\chi) \exp j \left(\omega t - k_{\text{ra}} z \frac{n_1}{n_2} \sin \alpha \right) \tag{4.67}$$

where

$$\chi = \frac{1}{k_{\text{ra}} (\sin^2 \gamma - 1)^{1/2}} = \frac{1}{k_{\text{ra}} [(n_1^2 / n_2^2) \sin^2 \alpha - 1]^{1/2}} \tag{4.68}$$

or

$$\chi = \frac{\lambda_2 n_2}{2\pi (n_1^2 \sin^2 \alpha - n_2^2)^{1/2}}. \tag{4.69}$$

The term λ_2 is the wavelength of the wave in medium 2. This is better expressed in terms of the free space wavelength λ_0 using

$$f = \frac{c}{\lambda_0}. \tag{4.70}$$

Similarly for medium 2,

$$f = \frac{v_2}{\lambda_2} = \frac{c/n_2}{\lambda_2}. \tag{4.71}$$

Combining equations (4.70) and (4.71) gives

$$\frac{c}{\lambda_0} = \frac{c/n_2}{\lambda_2} \tag{4.72}$$

so

$$\lambda_2 n_2 = \lambda_0. \tag{4.73}$$

Equation (4.69) can then be expressed as

$$\chi = \frac{\lambda_0}{2\pi(n_1^2 \sin^2 \alpha - n_2^2)^{1/2}}. \tag{4.74}$$

The value of χ is a measure of the depth of penetration of the electromagnetic fields into the forbidden region.

If this derivation is carried out separately for E and H, including the boundary conditions for the normal components to the boundary, a phase shift of $\pi/2$ appears between the two waves. Now the Poynting vector N for energy flow is given by $N = E \times H$. Consequently because of the phase shift there is no net flow of energy into medium 2. This protruding wave is termed *evanescent*.

4.2.2   Degradation of transmission in optical fibres

Currently, the transmission limit for an optical fibre telecommunications link between repeater stations is in excess of 100 km. This achievement has been made possible by the development of

- high quality manufacturing techniques and processes, and
- the materials from which the core and cladding are manufactured.

Transmission performance is degraded by attenuation and dispersion.

Attenuation

There are three mechanisms by which the intensity of the propagated light can be reduced.

Absorption

The fibre material can absorb photons. The electrons in the molecules from which the fibre is made have discrete energy levels. An electron can move to a higher energy state by absorbing a quantum of energy. Thus, if a photon having the same energy as that needed for the transition passes through the region of the molecule then it may be absorbed. This energy puts the molecule into an *excited* state. The energy is released some short time later as the molecule reverts to its *ground* state. The overall effect is to increase the thermal energy of the material. These electron energy levels in silicon-based fibres correspond to photons in the ultraviolet region.

In addition, the molecules of the fibre material may vibrate about their lattice points. The kinetic energies of these vibrations also have discrete

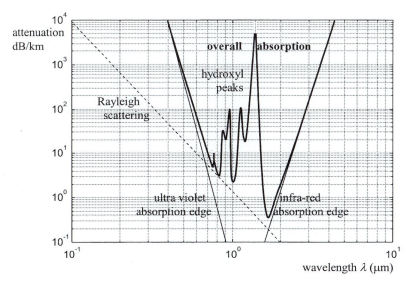

Figure 4.5. Representation of the absorption spectra for a silicon-based optical fibre.

values. Again, if a photon is in the vicinity, having the same quantum of energy as needed for an increase in the kinetic energy, then it may be absorbed. In this process the energies correspond to photons in the infrared region of 8–12 μm.

The absorption spectra in figure 4.5 illustrate these two processes. A minimum occurs at around 1.550 μm providing a low absorption window. The theoretically estimated attenuation level in silicon-based fibres has been virtually reached at 0.2 dB km^{-1} (Ohashi *et al* 1990).

The biggest absorption losses can come from minute traces of metals such as iron and copper in the glass. These may be present in the raw materials or even be introduced during the manufacturing process. Modern processing techniques have virtually eliminated this problem. However, it has proved very difficult to overcome the effects of the presence of the radical –OH. Also, if the fibre is exposed to hydrogen gas during its working life the number of –OH radicals can increase, causing an increase in attenuation. Attenuation peaks occur at wavelengths of 0.725, 0.825, 0.875, 0.950, 1.380 and 2.73 μm. See Keck *et al* (1973).

Exposure of the fibre to ionizing radiation or ultraviolet light is known to cause an increase in absorption. Ions produced in the glass act as absorption centres. It may be possible to employ these phenomena to create sensors.

Mode leakage

Attenuation may occur due to light leaking from the core into the cladding. As described in a previous section, light entering within the cone of acceptance and central to the axis will totally internally reflect. However, non-axial rays called *skew* rays may enter and propagate by reflections at just below the critical angle. They may travel hundreds of metres in a helical manner but are ultimately lost. When developing optical fibre sensors this mode of transmission may need to be accounted for.

If the fibre is bent, rays will strike the inner wall at less than the critical angle and so leak away. Manufacturers generally specify a minimum bend radius for their fibres that must be observed during installation to minimize losses. Microbends can occur during manufacture due to mechanical irregularities in the core-cladding interface.

Scattering

During the manufacturing process, as the fibre solidifies, minute local irregularities in refractive index occur. These are due to fluctuations in density and the concentrations of any dopants present. These irregularities are of the order of a tenth of the wavelength of the light being transmitted. The resulting effect is to cause light to be scattered out of the fibre. According to Rayleigh's law, scattering is proportional to λ^{-4}. So short wavelengths suffer significantly more scattering. For the currently available high-performance fibres this is the most significant loss mechanism. For silicon-based fibres it causes an attenuation of 1.7 dB per km at 820 nm.

In summary, the above attenuation mechanisms can be described using

$$I = I_0 \exp(-\alpha x) \tag{4.75}$$

where $\alpha = a_{\text{atomic}} + a_{\text{molecular}} + a_{\text{bending}} + a_{\text{scatter}}$.

Optical fibre joints and terminations also cause scattering and absorption. When a wave passes through a refractive index boundary some reflection occurs. For normal incidence this *Fresnel* reflection causes a 4% loss in transmission. This amounts to 8% or 0.35 dB for a joint. Optical coupling liquid or cement can be used for index matching. However, to achieve the lowest losses, fusing the components together whilst clamped in a special alignment jig is the standard approach. Losses are reduced to between 0.1 and 0.3 dB. Demountable connectors are available. These are particularly useful for research but require a high degree of mechanical precision for best results. Lenses can also be useful for coupling since they reduce the effects of axial misalignment and dust. Their disadvantage is that angular misalignment becomes more significant. Losses of around 1 dB per connector are typical.

Dispersion

Sir Isaac Newton is reputed to have discovered that a prism could disperse white light into its constituent colours and a further prism recombine the colours to restore the white light. This illustrates that refractive index is wavelength dependent. It follows that this is because the speed of an electro-magnetic wave within a medium is wavelength dependent.

Consequently, because no practical light source can be of a single wave-length, if a pulse of light is injected into an optical fibre the pulse length will increase during transmission. This is referred to as *chromatic* (colour) *dispersion*. It occurs within any mode so is classed as an *intramodal dispersion*. The group velocity of the pulse is dependent on the second derivative of the refractive index $d^2n/d\lambda^2$. Optical fibre engineers sometimes refer to materials in which this term is non-zero as showing *material dispersion*; see Senior (1985) page 81. A further, less significant, form of dispersion can occur in optical fibres due to the wavelength dependence of the angle a ray in a given mode makes with the fibre axis. This is sometimes referred to as *waveguide dispersion*. Again pulse broadening occurs but in multimode fibres the effect is negligible compared with material dispersion.

Pulse broadening during the propagation down a long fibre has similar consequences to intermodal dispersion described above. To keep this problem to a minimum, for long telecommunication links quasi-mono-chromatic light sources (laser diodes) are used.

Yamada *et al* (1980) found that the combined effect of material and waveguide dispersions is a minimum at a wavelength of 1.32 µm in a single mode fibre.

4.2.3   Manufacturing techniques

Early techniques used high quality glasses of the type used in lens manufac-ture. For a number of years the *rod and tube* method was used. In this process the core material, in the form of a rod, is inserted into a tube made of the cladding material. The two are heated to form a rod called the preform, which is then drawn to form the thin fibre. This technique is still employed for producing specialist fibres having a refractive index profile that is not symmetrical. Multiple rods of differing diameter and refractive index are placed within the tube of cladding material before being heated to produce the preform. Monomode fibres made in this way can preserve the orientation of polarization during light propagation. Another early development was the *double crucible* technique. Here two heated, concentric, platinum crucibles containing the molten core and cladding glasses share a common drain from which the preform is delivered. The advantage of this process is that it is continuous if the glasses are regularly replenished.

The fibre drawing system is common to all of the manufacturing techni-ques that have been developed. It is a high precision process. The preform is heated in the extrusion oven and the diameter of the drawn fibre output is monitored. A negative feedback system controls the draw speed. The fibre is also coated with its outer jacket before being wound on to the take-up reel. Proof testing of the finished product may also be carried out by applying a known tension via the take-up reel mechanism.

More recent fibre manufacturing procedures involve various forms of chemical vapour deposition techniques derived from the semiconductor industry. In the modified chemical vapour deposition process (MCVD), gas-phase reagents such as germania-tetrachloride and oxygen are burned together inside an externally heated silica glass tube. The resulting powder, under the influence of the heating, diffuses into the glass inner wall. The tube is later collapsed to produce the preform. By varying the types and concentrations of the reagents both core and cladding regions can be formed as well as graded refractive index profiles. A similar technique is the outside chemical vapour deposition process (OCVD). Here a rotating substrate or *bait* rod is used on which to deposit the materials produced by the MCVD method. The core materials are deposited first, followed by the cladding. The bait rod is removed and the glass layers collapsed to provide the solid preform. A refinement of this technique enables continuous rather than batch manufacture. Here the rotating bait rod is drawn through the burner region. The burner is mounted around the rod so as to produce the coating powders axially.

Table 4.1. Comparison of telemetry lines.

Characteristic	Twisted metallic pair	Coaxial cable	Optical fibre
Attenuation (dB/km)	3	2	1 (glass) 150 (polymer)
Distance-bandwidth product (MHz-km)	1	10^2	10^5
Repeater spacing (km)	1	5	100
System cost	Low	Medium	Medium
Lifetime (years)		Similar life spans of 25–40	
Crosstalk	High	Low	None
Noise immunity	Low	Medium	High
Electrical isolation	None	None	Total
Weight and size	High	High	Low
Termination technique	Solder	Solder with standard connections	Splicing by fusion of well-aligned ends
Fabrication control	Loose	Moderate	Precise

4.2.4   Comparison of optical fibres with electrical lines

Table 4.1 compares various transmission media characteristics for the common telemetry lines.

4.3   Extrinsic optical fibre sensors

Extrinsic optical fibre sensors have some external artefact connected to the fibre that is modified by the measurand. The fibre takes no part in the sensing. The optical fibre merely acts as the telemetry link and, on some occasions, as the power line. The majority of the optical fibre sensors developed are of this type. Udd (1991) gives a detailed catalogue of all of the types of optical fibre sensors. Some of the reported sensing processes used in extrinsic systems are described below.

4.3.1   Displacement sensors

A simple optical switch can be made using a shutter to interrupt the beam (figure 4.6) (see Jones 1985). S and D are the light source and detector fibres. This has been developed further to provide a continuous positional sensor using a Moiré-type fringe grating. The Moiré grating is moved across the gap between the faces of the transmit and receive fibres and the fringes generated are subsequently counted.

   A spherical lens has been used to move the light beam across the front of the fibre (see Dotsch *et al* 1983). The received intensity varies with lens displacement (figure 4.7). Two receiving fibres mounted side by side have been used to provide a differential change in light signals as the lens (or source fibre) moves across. This overcomes the problem of variation in source brightness.

   In section 4.2.1 above it was shown that the light entering at the front face of an optical fibre was only able to remain within the fibre if the angle the ray made to the fibre axis was within the cone of acceptance. As described previously, this also means that a transmitting fibre emits light within the

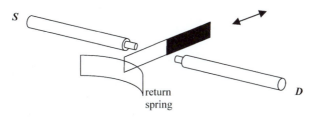

Figure 4.6. Binary optical switch.

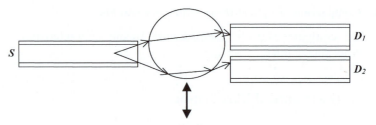

Figure 4.7. Spherical lens displacement sensor.

cone. Changing the separation of the fibres results in a change in the intensity of the received light (figure 4.8).

A strain gauge system and a hydrophone have been developed using this configuration. Kissinger (1967) filed a patent for a displacement sensor using this technique with a bundle of fibres in reflection mode. Frank (1966) filed a patent for a pressure sensor using just two fibres. The general arrangement for the transmit and receive beams for a two-fibre system is shown in figure 4.9.

With the reflector set at a distance of less than x_{min} from the front faces of the fibres the collecting fibre receives no light. This condition can be related to the radii r_S, r_D and the fibre separation y,

$$y - r_S - r_D = 2x_{min} \tan \theta_A. \tag{4.76}$$

If the reflector is moved so that the separation increases beyond x_{min} the intensity of the light received increases in a near linear manner and becomes a maximum when

$$y - r_S + r_D = 2x_{max} \tan \theta_A. \tag{4.77}$$

Increasing the separation further would cause a decrease in the received intensity tending towards an inverse square law relationship. The range x_{min} to x_{max} is set by the choice of r_S, r_D and their separation y. The shape of the transfer function was developed by Jones and Philp (1983). Figure 4.10 shows the received power for two different receiving fibres set at different distances from the reflector.

Jones and Philp (1983) have developed this concept further using three fibres. By the judicious choice of radii and separation one receiver fibre can

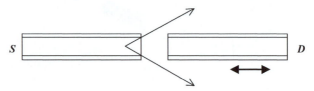

Figure 4.8. Fibre displacement sensor.

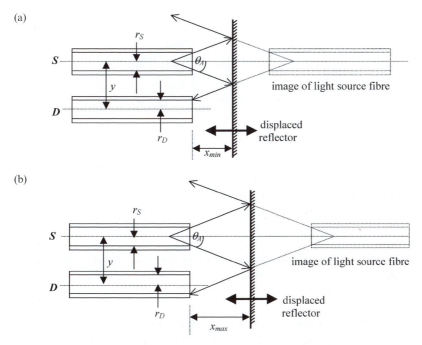

Figure 4.9. The reflectance mode displacement sensor: (a) minimum physical displacement. (b) Maximum physical displacement.

function as a reference, with the other providing the data (figure 4.11). The signal fibre gives a near linear response over the working range whereas the reference fibre is of almost constant output as indicated in figure 4.10. In this way any variation in light source performance can be accounted for.

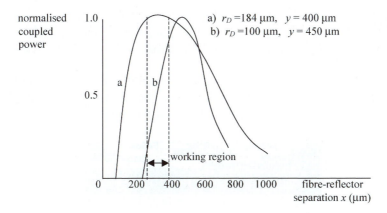

Figure 4.10. Received power as a function of displacement for two different geometries.

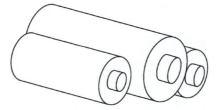

Figure 4.11. Stepped, three-fibre configuration.

This sensor was designed to measure pressure. Generally devices that measure small displacement can be adapted for the measurement of pressure, temperature, strain, vibration and acceleration.

4.3.2   Positional encoders

Precision translational or rotational position sensors can be made using a transparent encoded plate or disc. The Gray code described in section 5.3.1 and table 5.1 is used to minimize reading errors. The binary pattern is usually made photographically. For an 8-bit system, giving a resolution of 1 part in 256, eight optical fibres are required. Figure 4.12 shows a translational encoder plate covering the range 0–15 for an 8-bit system.

Multiplexing systems have been developed that only have a single transmit and receive fibre pair. This can be done by using a white light source separated into colours by using either filters, a diffraction grating or a prism. This is frequency-domain multiplexing although in fibre optics it is usually referred to as wavelength-division multiplexing. Figure 4.13(a) illustrates a prism configuration.

Alternatively, time-domain multiplexing can be used. Here short pulses of monochromatic light are transmitted into the system. Each receive fibre is made to be of different length so that a time delay is created between received data pulses. Figure 4.13(b) gives a schematic diagram. A length of fibre 0.2 m long corresponds to a delay of about 1 ns. Each fibre would need to be at least this much longer than its neighbour to ensure separation of pulses in the telemetry link.

The disadvantages are that the wavelength multiplexing technique requires precision optical engineering whereas time-domain multiplexing requires sophisticated, high-speed, electronic circuitry.

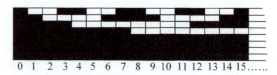

0 1 2 3 4 5 6 7 8 9 10 11 12 13 14 15......

Figure 4.12. Gray code positional encoder.

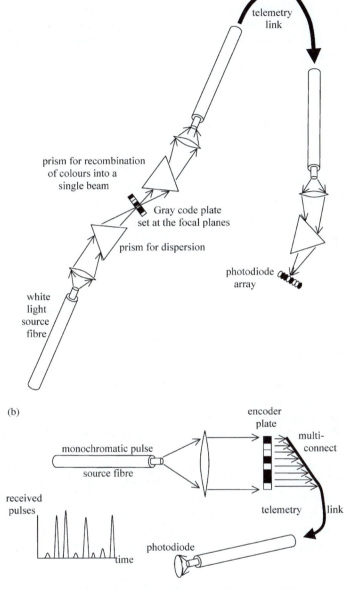

Figure 4.13. (a) Multiplexing by wavelength. (b) Time division multiplexing.

Diffraction gratings have also been successfully used to monitor displacement using either intensity variation or colour change. Prototypes have been developed for measuring displacement and pressure and for hydrophones—see chapter 7 of Udd (1991).

4.3.3   Absorption sensors

A number of different sensor types can be constructed based on light absorption. An input and an output fibre are required. Single-ended cell structures are the preferred physical design as this produces a more compact unit suitable for use as a probe.

The opacity of some liquid crystals is temperature dependent. Using an arrangement such as shown in figure 4.14 provides a simple temperature sensor. Unfortunately liquid crystals are only stable at moderate temperatures so the useful range is limited. This arrangement is best used as a temperature switch since the opacity–temperature relationship is nonlinear and of short span. A sandwich of crystals, each switching at a different temperature, would provide a thermometer, albeit with a stepped response.

The absorption of light passing through a gas or a solution can be used as a means of measuring its concentration. The absorption is described by the Lambert–Beer law,

$$\text{absorptivity} = \log_{10}\left(\frac{I_0}{I}\right) = \varepsilon c x \qquad (4.78)$$

where I_0 is the intensity of light received with no solute in solution, I is the intensity of light received with solute present, ε is the absorption coefficient of the species expressed in square metres, x is the distance between the transmit and receive fibres path in metres, and c is the concentration of the solution in moles per cubic metre.

Re-arranging equation (4.78) gives

$$I = I_0 10^{-\varepsilon c x} \qquad (4.79)$$

and assumes parallel light passing through the test cell (figure 4.15). The dimensions of the test cell and light source intensity need to be chosen for a particular application. GRINrod lenses could be used to provide a compact unit.

The dynamic range of the received light should be about two orders of magnitude. At the low end of the range the output signal needs to be greater than the noise level. It is common in this type of sensor to use two different wavelengths, one sensitive to the species to be measured and one not. In this way variation in the performance of the light source, or the presence of a species other than the one under test, can be accounted for.

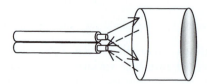

Figure 4.14. Single-ended absorption sensor using a mirror.

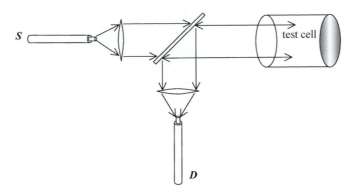

Figure 4.15. A test cell interrogated with parallel light.

If the chemical species itself is not a good absorber it is sometimes possible to use a reagent whose absorptivity changes with the concentration of the test species. An example of this is the use of a pH sensitive indicator such as phenol red for measuring acidity levels; see Dakin and Culshaw (1988). Reagent colour change can also be used although then the optical system becomes more involved, requiring spectral analysis.

In applications where a reagent is required it may be preferable to have the reagent absorbed in a membrane that is permeable to the species under test. One approach is to have an optically white (meaning 100% reflectivity over all relevant wavelengths) diffuse reflector (figure 4.16) (see Grattan and Meggitt 1995). The membrane is, in-effect, a coating on the white surface.

The Kubelka–Munk relationship, equation (4.80), is used to relate the reflectance with the concentration of the test species and assumes a semi-infinite scattering medium,

$$F(R) = \frac{(1 - R^2)}{2R} = \frac{\varepsilon c}{S} \qquad (4.80)$$

where R is the reflectance, S is a scattering coefficient and $F(R)$ is the Kubelka–Munk or *remission* function. This nonlinear relationship between

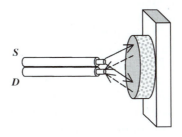

Figure 4.16. Absorption sensor using diffuse reflectance.

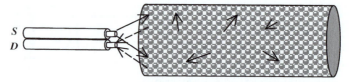

Figure 4.17. Absorption sensor using back-scatter with inert particles.

c and R means extensive characterization procedures are required otherwise interpolation between calibration points would yield high errors.

Finally, a common approach in absorption sensors is to use light scattering within the test cell rather than some form of reflector to increase performance (figure 4.17). Although backscattering is not as efficient as a reflector at returning light to the detector fibre, the level of source light can easily be increased to compensate. Scattering does ensure thorough irradiation of the test cell. Particles of about 1 μm are used. A commonly used source of such particles is polystyrene latex. It is also possible to mount the reagent on scattering particles if it is not practical for it to be in solution. Polyacrylamide microspheres have been used to support phenol red to produce a pH probe for physiological applications; see Kirkbright *et al* (1984).

Each of the test cells in figures 4.14–4.17 could be constructed with a selective membrane that permits the entry of the test species but retains the reagents within the cell. Cellulose dialysis tubing is used for aqueous solutions whereas silicone rubber is more appropriate for gas systems.

Multiple reagent probes have been produced that detect specific species through two or more chemical reactions. The only limit on the development of these probes is in obtaining appropriate reagents that, ideally, are reusable. The name *optrode* is becoming popular for optical sensor probes.

4.3.4 Luminescence sensors

Luminescence is the emission of light from a previously excited molecule as it reverts to a lower energy state. The intensity of the emitted light declines exponentially with time. It can be shown that, for a large sample, the decay time constant τ is equal to the average time a molecule exists in its excited state. This is usually referred to as the *lifetime*. There are two classes of luminescence, *fluorescence* and *phosphorescence*. Fluorescence refers to emission where lifetimes are short, of the order of 10^{-9} s, and occurs in various solids, liquids and gases. Phosphorescence only occurs in solids and, because the transition to the lower energy state is classed as *forbidden*, the light emission may be over many minutes. Emission intensities are much lower with phosphorescence and generally require more sensitive detectors or time integration of the signal or both. For a text on fluorescence see Lakowicz (1983).

Fluorescence sensors

Light is admitted from the transmitting fibre and is absorbed by the medium according to the absorption relationship given above in equation (4.79). The response of the fluorescent material is monitored. There are several mechanisms by which fluorescence may occur; see Scheggi and Baldini (1993).

1. The species to be measured is itself a fluorescent material.
2. Molecules of the test species are made to attach to molecules of a fluorescent reagent called a *fluorophore* making them fluorescent. This type of procedure is sometimes referred to as *labelling*.
3. The species reacts with a reagent to form a new complex or compound that is fluorescent.
4. The species interacts with a fluorophore so that the fluorescence is modified in intensity or lifetime.

One form of this latter mechanism is the phenomenon of quenching. The test species has the effect of either preventing the fluorophore from absorbing the light transmitted to it (*static quenching*) or altering the emission process of the excited fluorophore (*dynamic quenching*). With either phenomenon the fluorescent intensity is reduced. For both types of quenching the relationship between the intensity of the fluorescence I and the quencher Q present in concentration $[Q]$ is given by

$$\frac{I}{I_0} = \frac{1}{1 + K[Q]} \qquad (4.81)$$

where I_0 is the fluorescent intensity in the absence of the quencher. For static quenching K is the dissociation constant of the reaction between the fluorophore and the test species. For dynamic quenching it is the Stern–Volmer constant which involves the lifetime of the uncombined fluorophore and the bimolecular quenching constant.

For dynamically quenched fluorophores measurement of the lifetime may be an alternative to the measurement of intensity. The lifetime of the fluorophore without the presence of the test species is a reference value, τ_0. The lifetime is reduced to τ because the excited state of the fluorophore is depopulating non-radiatively in addition to the normal, light-emitting process. The lifetimes are related by a similar expression to equation (4.81),

$$\frac{\tau_0}{\tau} - 1 = K_{SV}[Q] \qquad (4.82)$$

where K_{SV} is the Stern–Volmer constant; see Dakin and Culshaw (1997).

An advantage of this technique is that the intensity of the light input I_0 is not part of the determination. This makes the system immune to variations in the performance of the light source or to tarnishing of the optical components.

With luminescence techniques the emitted light is of a different wavelength from that of the exciter. This can give up to $\times 10$ greater sensitivity compared with the straightforward absorption technique. However, absorption techniques can often be more immune to the presence of other species.

Commercial devices have become available. An interesting example is in blood analysis where partial pressures of O_2 and CO_2 can be monitored as well as pH.

Phosphorescence sensors

Certain rare earth compounds phosphoresce with an exponential decline that is related to their temperature. This phenomenon has been used to produce a temperature sensor. The structure of the device is simply a bi-directional fibre having the active material coated over the front end. This end is in thermodynamic equilibrium with the environment to be measured. These materials function in the range -200 to $+400\,°C$. The light emission L can be expressed as

$$L = L_0 \exp\left(\frac{-t}{\tau}\right) \tag{4.83}$$

where L_0 is the initial phosphorescent intensity as the energizing light pulse is switched off. The value of the time constant τ is measured and from calibration tables the temperature is determined. The value of τ is reduced as the temperature is increased, typically ranging from $5\,s$ at $-200\,°C$, to about $1\,s$ at $+400\,°C$. As with the fluorescence case, because the time constant is being measured to determine the value of the measurand, variation in source intensity or attenuation in the system does not affect calibration. A commercial device is available from the Luxtron company.

4.3.5 Total internal reflection-based sensors

A liquid level sensor has been constructed using $90°$ prisms (figure 4.18). The angle of incidence is $45°$, which is larger than the critical angle for a glass to

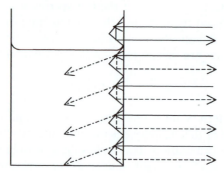

Figure 4.18. Liquid level sensor.

air interface. However, if the liquid is allowed to surround the prism light will not totally reflect and the detector will have a much-reduced signal. An array of such sensors down the side of the tank provides discrete level measurement (Jones 1981).

A more sophisticated application of the total internal reflection is in the use of the evanescent field. This was described toward the end of section 4.2.1. Following equations (4.67) and (4.74), the amplitude of the electric field component can be written as

$$E = E_0 \exp(-x/\chi) \tag{4.84}$$

where

$$\chi = \frac{\lambda_0}{2\pi(n_1^2 \sin^2 \alpha - n_2^2)^{1/2}} \tag{4.74}$$

and α is the angle of incidence, which must be greater than the critical angle, and λ_0 is the free space wavelength.

Equation (4.74), using typical values of n_1, n_2 and θ, gives χ as about one wavelength. There is no energy transfer as the H and E vectors are out of phase. The result is that energy flows in during one half-cycle but is returned in the next.

To make use of this phenomenon in sensing it is necessary for power to be coupled out of the first medium via the evanescent wave. Thus the coupling component has to come within about a wavelength of the first medium. When this occurs the totally internally reflected wave is reduced in power. There are two modes of construction for evanescent wave sensors.

Front end sensor

Here the evanescent wave penetrates into the material placed within a wavelength of the obliquely cut front end of the fibre. Quantitative chemical analysis has been achieved by having the test solution in contact with the end of the fibre. Changes in the concentration of the solution alter the value of its refractive index which affects χ.

The measurement of temperature and strain has been reported using two fibres. One fibre is fixed and the other moves in response to the measurand so that the coupling of the light into the second fibre is affected (figure 4.19). This effect is sometimes referred to as frustrated total internal reflection. Spillman and McMahon (1980) have used this approach to produce a hydrophone.

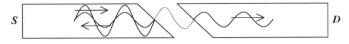

Figure 4.19. Frustrated total internal reflection.

Side-coupling sensor

In this mode of operation the cladding is removed from a section of the fibre allowing the evanescent wave to form around the sides. A similar fibre is mounted alongside in parallel, about a wavelength away, in order to receive the wave. By allowing the measurand to alter the separation of the fibres a temperature or strain sensor can be produced similar to the front-end system above.

A more sophisticated alternative is to coat the exposed fibre with some chemical or biological reagent whose refractive index will vary with the concentration of the species to be measured. A number of patents have been established for biological sensors using this technique. The attractive features are as follows.

- The evanescent wave is so localized that only the behaviour of the bonded reagent affects the sensor and not the general solution, thus providing a high selectivity.
- Using the disposable capillary structure shown in figure 4.20 means that a fixed volume of the analyte (test solution) is drawn inside. The sensitized

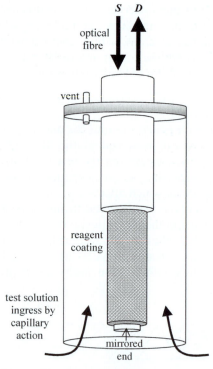

Figure 4.20. A disposable, capillary-fill, evanescent wave sensor.

region contains an appropriate reagent dye for monitoring the level of some chemical species or an antibody to monitor the level of the corresponding antigen. In either case the evanescent wave is variably absorbed by the response of the reagent and absorption spectroscopy can be performed. Alternatively, the reagent may be a fluorescent material, enabling fluorescent spectroscopic analysis of the reflected signal.

A number of evanescent wave sensors have been built in hybrid, miniature form on a substrate. A typical structure involves a Mach–Zehnder interferometer described in section 4.4.1. An example is the hydrogen sensor reported by Nishizawa *et al* 1986. These and similar structures are sometimes referred to as *integrated optic* devices or *optical integrated circuits* (Nishihara 1989).

4.3.6 Pyrometers

Pyrometry is the measurement of the temperatures of incandescent objects from their colour. The emission spectrum of a perfect black-body is given in figure 3.13. The traditional pyrometer is a remote, hand-held device that produces an image of the hot object on a translucent screen. This is viewed through the eyepiece alongside a set of coloured templates that are calibrated for temperature. The user obtains the best match of colour and then reads the temperature from the template. Because of the variation of emissivity in objects which are not perfect black-bodies the templates are calibrated for a particular application, such as molten iron in a blast-furnace.

To produce an optical fibre pyrometer a material capable of withstanding furnace temperatures is needed for the probe section. A single crystal of sapphire has proved suitable. This is coated with a high melting-point metal so as to provide a shield to any visible light and also act as the black-body cavity walls (figure 4.21). Iridium is a good choice, despite being hard to work, as it is the most resistant material to chemical attack. The sapphire is optically coupled to a silica fibre outside the hot environment.

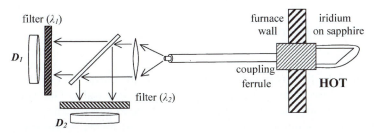

Figure 4.21. An optical fibre pyrometer.

Since the black-body cavity is integral to the instrument a correction for the non-perfect emissivity can be incorporated. Also, coupling and absorption losses in the system can be accounted for by using the relative intensities of two different parts of the spectrum using narrowband filters. Resolutions of 0.5 °C are possible with a tolerance of ±0.0025% in the range 500–2000 °C. A commercially available optical fibre pyrometer has been produced by Luxtron.

4.4   Intrinsic optical fibre sensors

A strict definition of intrinsic sensing is that the measurand modifies some parameter or property of the optical fibre. Measurement occurs within the fibre. The benefits of intrinsic optical fibres in addition to those of the extrinsic type are

- greater geometrical versatility,
- wider dynamic range,
- much higher sensitivity,
- high frequency response.

The disadvantages of intrinsic sensors compared to the extrinsic types are generally

- greater complexity,
- more sophisticated signal recovery and processing is required,
- greater precision and stability of optical components is required,
- a more highly specified and stable light source is needed,
- greater sensitivity to ambient parameters reduces selectivity and causes drift, making the measurement of steady or slowly varying measurands difficult.

The optical phenomena used for measurement are

- attenuation or phase shift due to physical distortion of the fibre,
- rotation of polarization,
- scattering.

4.4.1   Physical distortion

Microbending

In section 4.2.1 the propagation of light down an optical fibre was described and the need for total internal reflection at the fibre walls. The calculations assumed an axially straight fibre. With microbend sensors the curvature of the fibre axis is altered by the measurand so that losses in transmission do occur. The arrangement is shown in figure 4.22.

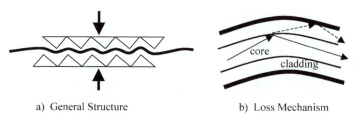

a) General Structure   b) Loss Mechanism

Figure 4.22. A microbend sensor.

The microbend technique has been used to produce position, force, pressure, strain, acceleration and vibration sensors. Nonlinearities of less than 1% have been reported.

4.4.2 Change in optical-path length

In section 4.2.1 the wave equation was written as

$$Y = Y_0 \exp j(\omega t - \boldsymbol{k} \cdot \boldsymbol{s}). \tag{4.32}$$

The term $(\omega t - \boldsymbol{k} \cdot \boldsymbol{s})$ completely describes the phase of the wave at any position and instant in time. For simplicity, assuming that the wave travels axially in the z direction, the phase term can be re-written as $(\omega t - kz)$, where k is the free-space wave number $2\pi/\lambda_0$. During transmission through some material this free-space wavelength is shortened since $v_1 < c$.

$$c = f\lambda_0 \tag{4.85}$$

$$v_1 = f\lambda_1 \tag{4.86}$$

dividing

$$\frac{c}{v_1} = n_1 = \frac{\lambda_0}{\lambda_1}. \tag{4.87}$$

Consequently, the phase term in the medium is expressed as $(\omega t - n_1 kz)$ where n_1 is the absolute refractive index. The term $n_1 z$, the physical path multiplied by the refractive index, is known as the *optical path length*.

If the wave were to divide, be sent along two different paths, and then recombine there would be some phase difference ϕ where

$$\phi = n_1 kz_1 - n_2 kz_2. \tag{4.88}$$

If the two paths are within the same material,

$$\phi = nk(z_1 - z_2) = nk\delta z. \tag{4.89}$$

Since these phase shifts are fractions of the wavelength of light the potential exists for measuring tiny displacements or changes in refractive index. The construction of the sensor must be such that the measurand generates some change in path difference. These phase shifts are measured by interference

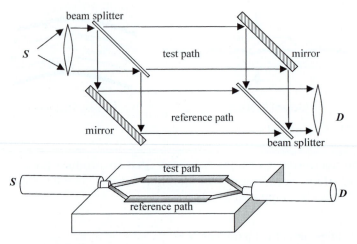

Figure 4.23. Mach–Zehnder interferometer.

effects using some form of interferometer. Unfortunately, the light intensity variation with path difference is not a linear function.

There are three historic designs of interferometer that have been adapted for optical fibre sensing. These are the Mach–Zehnder (figure 4.23), the Fabry–Pérot and the Michelson types and are illustrated below in original and optical fibre forms. The sensor element can be constructed on a substrate as an optical waveguide. This substrate-based technology is often referred to as integrated optics; see Göpel *et al* (1992). The most commonly used for optical fibre sensing is the Mach–Zehnder interferometer.

The Fabry–Pérot interferometer (figure 4.24) has the advantage that the light passes many times through the test region which is advantageous in some applications.

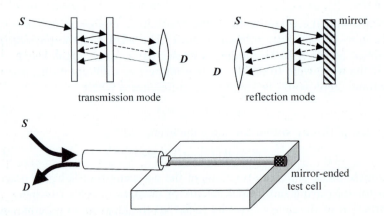

Figure 4.24. Fabry–Pérot interferometer.

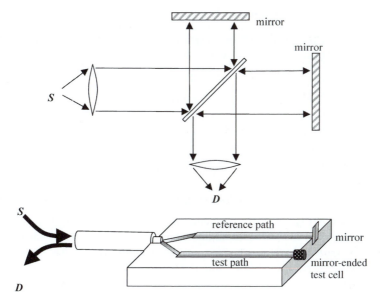

Figure 4.25. The Michelson interferometer.

The Michelson interferometer (figure 4.25) has some similarities to the Mach–Zehnder design. For a comparative analysis of the interferometer types see Udd (1991).

A significant problem faced by designers of interferometer-based sensors is that steady state or slowly varying measurands are difficult to measure accurately due to thermal drift causing changes in the optical path difference. Such drifting has a time scale of >0.1 s. Consequently, interferometric sensors are best suited to measurands varying at >10 Hz. A further difficulty lies in the measurement of lengths greater than one wavelength since identical interference intensities are reproduced throughout every 2π phase-shift. The zero setting also becomes ambiguous. Thus small, relative displacements can only be measured unless a more sophisticated auto-zeroing construction and additional signal analysis is employed.

Optical fibre interferometers have been developed to measure refractive index, acceleration, electric and acoustic fields, electric current, temperature, strain and pressure. Refractive index can be used for chemical analysis measured either at the end of the fibre in a Michelson or Fabry–Pérot integrated optic structure, or through the side of the fibre via the evanescent-wave using a Mach–Zehnder integrated optic interferometer. Described below are three devices that have achieved some commercial success.

Magnetostriction magnetometer

Ferro-magnetic materials exhibit magnetostriction and below saturation the longitudinal strain produced e, which is $\delta z/z$, generally goes as

$$e = CH^2 \tag{4.90}$$

where C is a material constant; see Livingston (1982).

A magnetometer can be constructed by directly bonding a material having magnetostrictive properties to the surface of an optical fibre. In the presence of a magnetic field the fibre will either be stretched or compressed axially, thus causing a change in the phase of the light transmitted. From equation (4.89),

$$\phi = nk\eta\xi zCH^2 \tag{4.91}$$

where η is the strain transfer efficiency and ξ is a correction factor for the change occurring in k due to axial strain in the fibre core. Driving the system at mechanical resonance improves sensitivity. An appropriate interferometer structure is the Mach–Zehnder type. Dagenais *et al* (1988) reported a resolution of $20\,\mathrm{pT}\sqrt{\mathrm{Hz}}$ at $1\,\mathrm{Hz}$ operating at a mechanical resonance of $26.5\,\mathrm{kHz}$.

Hydrophone

Rapid developments in hydrophones have taken place since the 1970s driven particularly by the military services. Optical fibre based hydrophones are attractive due to their structural versatility and relatively low cost. Noise levels of at least $10\,\mathrm{dB}$ below ambient underwater noise have been achieved with sensitivities of $-300\,\mathrm{dB}\,\mu\mathrm{Pa}^{-1}$.

There are two approaches to construction. The fibre is either simply given a flexible coating and mounted perpendicular to the vibration axis as in figure 4.26(a) or wound around a compliant mandrel to increase sensitivity (figure 4.25(b)).

Sagnac effect

A great deal of effort has been expended over the past thirty years on the development of an optical fibre gyroscope based on the Sagnac effect reported in 1913. See Post (1967) for an account of the Sagnac effect.

(a)   (b)

Figure 4.26. Optical fibre hydrophones.

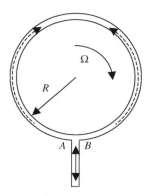

Figure 4.27. The Sagnac effect.

An optical arrangement for observation of the Sagnac effect is illustrated in figure 4.27. The optical medium is a ring of radius R rotating at some angular velocity Ω together with the detector. The time of travel from A to B is longer than that from B to A due to the rotation. Consequently, if a beam of light is split into two and one is made to propagate in the direction of rotation while the other is in the counter direction, a phase difference will occur. This phase difference is effectively due to the difference in path length. A simple classical approach assumes two counter-rotating light paths.

The time Δt taken for the light to travel one circuit around the ring, when the ring is stationary, is

$$\Delta t = \frac{2\pi R}{v} \tag{4.92}$$

where v is the speed of light in the glass.

If the ring rotates then during Δt it will have turned through $\Omega \Delta t$ radians. Consequently, the path AB will have increased by $R\Omega \Delta t$ and for BA reduced by $R\Omega \Delta t$. Therefore, the path difference is

$$\Delta z = 2R\Omega \Delta t = 2R\Omega \frac{2\pi R}{v} \tag{4.93}$$

$$\Delta z = 4A\frac{\Omega}{v} \tag{4.94}$$

where A is the area of the ring. The phase difference ϕ, from equation (4.89), is $nk \Delta z$,

$$\phi = nk4A\frac{\Omega}{v} \tag{4.95}$$

which would suggest the result

$$\phi = n^2 8\pi A \frac{\Omega}{c\lambda_0}. \tag{4.96}$$

However, Post (1967) reported that experimental observation by various workers yielded the result that, for the case where the propagation medium is co-moving with the interferometer, the presence of the medium has no effect on the interference fringes observed. So in equation (4.96) the refractive index is effectively unity giving

$$\phi = 8\pi A \frac{\Omega}{c\lambda_0}. \tag{4.97}$$

Post also reports that the Sagnac effect does not depend on the shape of the surface area and does not depend on the location of the centre of rotation.

From equation (4.97) the phase shift is directly proportional to Ω and hence the usefulness of the technique for inertial guidance systems. The best devices currently available can detect rotations as low as 0.00001 degrees per hour.

4.4.3   Rotation of polarization

It is useful first to review the phenomena of polarization. It was shown in section 4.2 that the electric and magnetic fields in an electromagnetic wave are orthogonal to each other and the direction of propagation (figure 4.1). As a convention the electric field direction is used to describe the orientation of polarization. A wave propagating with some fixed orientation of E is said to be *plane* polarized, the electric field lying in a plane through the axis of propagation. Photons are emitted from an ordinary light source having all possible orientations of E around the propagation axis. This light is described as unpolarized. The superposition of beams of polarized light was investigated by Fresnel and Arago (see Longhurst 1970). The three laws they discovered are summarized here to help appreciate the nature of polarized light.

1. Orthogonally, plane-polarized beams of light cannot produce interference effects.
2. Two plane-polarized beams of light can produce interference effects provided that the orientation of their planes of polarization is the same *and* that the beams were derived from the same plane-polarized source or the same plane of polarization of an unpolarized source.
3. Two plane-polarized beams of light derived from perpendicularly polarized components from unpolarized light *never* produce interference effects even when their planes of polarization are rotated so as to have the same orientation.

A number of techniques are available that can be used to separate or filter light waves according to their polarization.

Polarization by glass plate

Polarization can be achieved using an optical boundary. When an unpolarized beam crosses an optical boundary at an oblique angle partial polarization of the reflected and refracted waves takes place. The waves have polarizations normal to the plane of incidence and in the plane of incidence respectively. The amount of polarization that occurs is a function of the angle of incidence. The reflected wave is completely polarized when the angle of incidence is Brewster's angle. Brewster's angle is $\tan^{-1}(n_{\text{refraction}}/n_{\text{incidence}})$; see Bleaney and Bleaney (1965). This is sometimes made use of in lasers. Laser light is polarized but unless there is some mechanism in the lasing cavity, such as a glass plate inclined at Brewster's angle, the orientation of polarization randomly fluctuates with a timescale of the order of 1 ms.

Polarization by birefringence

One of the phenomena that can occur in optically active materials is that the speed of propagation of the wave is dependent on its polarization. This effectively means that the material has differing refractive indices for different orientations of polarization. This phenomenon is termed *birefringence*. Naturally occurring examples of birefringent materials are crystals of quartz and calcite (Iceland spar). If an unpolarized beam of light falls on a single crystal at the correct orientation two waves emerge, each plane polarized and whose planes of polarization are orthogonal. (It is easy to see why the original name for this phenomenon was *double refraction*.) To understand this effect it should be remembered that, as stated above, an unpolarized beam contains all possible orientations of polarization in equal measure. It is therefore possible to resolve the electromagnetic fields of a beam into two orthogonal axes in a plane perpendicular to the direction of propagation. If these axes are appropriately inclined to the optically asymmetric axes of the crystal then the two components of the beam will refract differently and so separate.

Polarization by selective absorption

Polarization can also be achieved using dichroic materials. Dichroism is the selective absorption of one of the orthogonal polarizations within an unpolarized light wave. The phenomenon, discovered by Biot in 1815, occurs in some minerals, the most well-known being tourmaline, and in some organic compounds. The development of polarizing film was pioneered by Land and first patented in 1932; see Land (1951) for a review of early developments. The Polaroid Corporation was formed by Land and Wheelwright in 1933. The ratio of the polarized components after passing through a polarizing film can be as high as 10^5.

Light propagating through optical fibre can suffer change of polarization due to various external influences. Such phenomena can be used to provide a sensing mechanism.

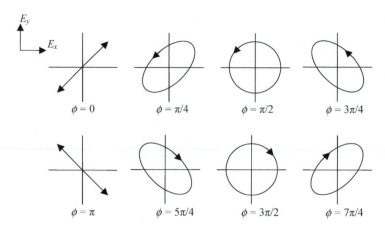

Figure 4.28. States of polarization (E_x leads E_y).

It is helpful to consider the various states of polarization that the output wave might develop due to birefringence. The presence of birefringence may be intrinsic to the glass or induced by the measurand. A plane-polarized wave input to the fibre can be resolved into two identical orthogonal components having planes of polarization, say E_x and E_y, travelling along a birefringent fibre in the z direction. Assuming that there is no attenuation, when the waves emerge into the detector system the electric field, formed from the vector addition of E_x and E_y, will have three possible forms (figure 4.28).

If the retardation of one wave relative to the other is zero or a multiple of π, the resultant electric field vector will be of fixed orientation in time. This state is termed *linear* polarization. (Linear polarization and plane polarization are in effect the same state but the use of the word linear groups the phenomenon with the other phenomena described next.) If the retardation of one wave relative to the other is an odd multiple of $\pi/2$ then the tip of the resultant electric field vector will trace a circle. This is termed *circular* polarization. All other retardations, and therefore the most common, cause the tip of the electric field vector to trace an ellipse hence the term *elliptical* polarization. Both of these vector rotations occur at the optical frequency and will not be seen visually. Also, if as stated above there is no attenuation the linearly polarized beam will emerge with the same angle of polarization as the plane wave input. If attenuation does occur, as well as affecting this angle, circular polarization will not be possible. Instead an ellipse will be generated having major and minor axes aligned to the x and y axes.

Clearly, for sensor use the intrinsic state of polarization within the fibre needs to be stable so that change caused by the measurand can be recovered. To achieve this the fibre must be monomode and have one of the following criteria.

1. Have no birefringence. This is difficult to achieve, requiring special materials and high precision manufacturing techniques.
2. Have high birefringence along the axis by deliberately creating stress during the manufacturing process. The differing refractive indices for the two orientations means that it is possible to arrange for one mode to be lost in the cladding so only one plane of polarization propagates. The cladding mode is stripped away at the receiver.
3. Have a tiny core of elliptical cross-section. Polarised light injected having its polarization aligned with the major axis of the ellipse will propagate without change of orientation.

Some of the polarizing phenomena that have been used in sensor design are described below.

The Faraday magneto-optic effect

In 1845 Michael Faraday discovered that, when plane-polarized light passes through glass subjected to a strong magnetic field aligned with the direction of propagation, the plane of polarization is tilted. This effect is as a result of the interaction of the magnetic fields of the light and electron orbitals. The angle of tilt θ is given by

$$\theta = V \oint \boldsymbol{H} \cdot d\boldsymbol{l} \qquad (4.97)$$

where $d\boldsymbol{l}$ is the closed loop path increment along the field line $\boldsymbol{H}$ and V is the Verdet constant for the level of the effect in a given material. V is wavelength dependent but temperature has little effect on the process. Its value is typically 5×10^{-6} rad A^{-1} for near infrared light in silica. Response times are much faster than in sensors involving magnetostriction. Bandwidths of 1 GHz have been achieved.

A most useful application of this technique has been the development of an optical fibre current sensor (figure 4.29). Pioneering work was carried out for the CEGB research centre by Rogers (1979) at Fawley power station in England. The sensor is formed by wrapping the optical fibre around the

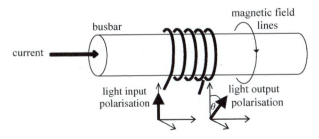

Figure 4.29. An optical fibre current sensor.

busbar. From Ampère's law,

$$I = \oint \mathbf{H} \cdot \mathrm{d}\mathbf{l} \tag{4.98}$$

so equation (4.97) becomes

$$\theta = nVI \tag{4.99}$$

for n turns of fibre. Since the size and position of the busbar is not part of the calculation, vibrations of the conductor are not important. However, vibration of the fibre was found to affect polarization.

Polarization can be measured using a Wollaston prism. This produces two orthogonal, linearly polarized output beams. The two beam intensities I_1, I_2 are measured and the current determined using the fractional relationship

$$\theta = K \frac{I_1 - I_2}{I_1 + I_2} \tag{4.100}$$

where K is a constant of the particular fibre. This use of a fractional intensity relationship overcomes problems of light source drift and path change.

Research has also been carried out by Siemens in Munich and Kema in the Netherlands.

Kerr effect

In 1875 Kerr discovered that when an electric field is applied to an isotropic, dielectric material it becomes birefringent. A plane-polarized light wave having its electric field aligned with the applied field is slowed by this field. In the Faraday magneto-optic effect described above it is the magnetic field from the light wave that interacts with the rotary motion of the electron orbitals. In the Kerr effect it is the light electric field that interacts with the linear component of the electron orbital motion. The Kerr effect is quantified by the change in refractive index caused and is dependent on the square of the electric field

$$\Delta n = K\lambda_0 E^2 \tag{4.101}$$

where K is the Kerr coefficient. The resulting phase difference ϕ suffered by the wave due to the applied field is determined using equation (4.89) giving

$$\phi = 2\pi K E^2 \, \mathrm{d}z. \tag{4.102}$$

The response time of the Kerr effect is of the order of 1 ps which has led to applications in high speed optical shuttering.

The measurement of electric field strength can be achieved by observation of changes in polarization. A plane-polarized wave inclined at 45° to the field to be measured is injected into the fibre and becomes elliptically polarized because of the slowing of its component aligned with the external field. The form of the ellipse is then analysed to measure the field strength.

Both current and voltage sensors were investigated by the CEGB. For voltage measurement the electric field must be applied transversely to the fibre. This was achieved by mounting the fibre in spiral form and sending a light pulse up to the high-voltage line and reflecting it back to the detector at ground potential. The state of polarization depends on the square of the electric field as indicated in equation (4.102). Because of this nonlinearity the voltage measuring device proved more difficult to realize than that of the current due to local changes in the electric field occurring across the fibre. This was due to dust, moisture and vibration. It was suggested that development of an appropriate crystalline fibre, having a linear response rather than squared, might overcome this problem. Alternatively, the use of polarization, optical-time-domain reflectometry, as described in section 4.5.2, could overcome the integration of the nonlinearity error.

The Kerr effect can also occur due to interaction with the electric field form another light wave within the dielectric. This is referred to as the *optical Kerr effect*.

The photoelastic effect

A material is said to have photoelastic properties if its refractive index can be made to alter by some applied stress σ or strain ε. Photoelasticity was discovered by Brewster in 1816. A wave having its plane of polarization in the same direction as the stress travels faster than a wave with the orthogonal polarization. Common materials exhibiting photoelasticity are gelatin, glass and polycarbonate. The phenomenon has been used for several decades by engineers involved in stress analysis. Usually a model is built of the structure to be investigated in polycarbonate or similar photoelastic material. White light is polarized, passed through the model and viewed through a further polarizer. Stresses in the photoelastic material are said to cause *retardation* between orthogonal polarizations of the components of the wave. This results in destructive interference of a particular wavelength. The observer sees secondary colours where the stresses are located. The compactness of these fringe patterns indicates the levels of stress.

The following paragraphs show how quantitative analysis of stress or strain can be achieved. Figure 4.30 illustrates a wave travelling in the z direction through a photoelastic material being stressed in the x direction. The retardation can be described by the relation

$$\text{retardation} = \text{stress} \times \text{stress optic coefficient} \times \text{optical path.}$$

Retardation is usually quoted as a physical length, such as 150 nm. This can be expressed as a phase shift,

$$\text{phase shift} = \text{stress} \times \text{stress optic coefficient} \times \text{optical path} \times \frac{2\pi}{\lambda}$$

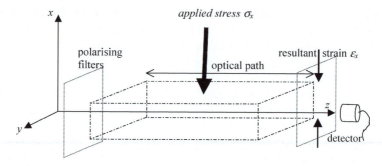

Figure 4.30. The photoelastic effect.

or

$$\phi = \sigma_x C_\sigma nz \frac{2\pi}{\lambda}. \tag{4.103}$$

In some applications working with strain is preferred. Retardation is then given as

retardation = strain × strain optic coefficient × optical path.

Equation (4.103) then becomes

$$\phi = \varepsilon_x C_\varepsilon nz \frac{2\pi}{\lambda}. \tag{4.104}$$

The usual arrangement is to have a polarizing filter placed in front of the photoelastic material. The plane of polarization is set at 45° to the x axis. In this way light from an unpolarized source is made to have equal components in the x and y directions. The emergent light passes through a further polarizer placed in its path and finally a detector simply measures the optical power.

Using equations for the electric and magnetic energy stored in an electromagnetic wave, it can be shown that the instantaneous rate of energy flow per unit area P is the cross product of E and H,

$$P = \tfrac{1}{2}E \times H. \tag{4.105}$$

P is known as the Poynting vector and its direction is that of the energy flow. The average rate of energy flow per unit area, usually referred to as the intensity, is determined from the rms values of E and H. For a plane wave the ratio of E to H is a useful relationship and has been called the *intrinsic impedance* of the medium Z_0. For a plane wave polarized in the x direction

$$Z_0 = \frac{E_x}{H_y} = \sqrt{\frac{\mu}{\varepsilon}}. \tag{4.106}$$

Combining equations (4.105) and (4.106) for a plane wave polarized in the x direction gives

$$P_x = \frac{1}{2}\sqrt{\frac{\varepsilon}{\mu}}E_x^2. \tag{4.107}$$

Generalizing, equation (4.107) indicates that for a wave of any plane of polarization the intensity is proportional to the square of the amplitude of the electric field in that plane.

Now, since the detector in figure 4.30 will be measuring the intensity resulting from the combination of two orthogonal components, the separate amplitudes must be combined before calculating the intensity. The variation in electric field for a wave polarized in the x-plane can be expressed as

$$E_x = E_{ax}\sin(\omega t + \phi) \tag{4.108}$$

where E_{ax} is the peak amplitude. A similar expression is used for E_y but without the phase shift ϕ.

Finally, if the second polarizer has its orientation set at some angle θ to the x axis, the resultant electric field will be

$$E_{\text{total}} = E_{ax}\sin(\omega t + \phi)\cos\theta + E_{ay}\sin\omega t\sin\theta. \tag{4.109}$$

But, as stated above, the instantaneous intensity measured by the detector will be related to the square of the resulting electric field, so squaring equation (4.109):

$$E_{\text{total}}^2 = [E_{ax}\sin(\omega t + \phi)\cos\theta + E_{ay}\sin\omega t\sin\theta]^2. \tag{4.110}$$

Averaging over one cycle and assuming $E_{ax} = E_{ay}$, equation (4.110) gives the mean as

$$\bar{E}_{\text{total}}^2 = \frac{E_{ax}^2}{2}(1 + \sin 2\theta\cos\phi). \tag{4.111}$$

The above derivation has been performed for two orthogonally polarized waves. But, as described following the diagram above, a wave plane-polarized at 45° to the x axis is used to generate the two orthogonal waves. If the intensity of this input wave is I_0 each orthogonal component will have an intensity $I_0/2$. (This is verified from resolving the input wave electric field E_0 into its x and y components; $E_x = E_y = E_0\cos 45$ and remembering that intensity goes as E^2.) So expressing equation (4.111) in terms of intensity gives the intensity at the detector as

$$I = \frac{I_0}{4}(1 + \sin 2\theta\cos\phi). \tag{4.112}$$

The usual practice is to set θ to $-45°$ so that for zero retardation I is zero. A further subtle adaptation is to pre-retard one of the orthogonal components by the use of a *quarter-wave plate*. Increasing the path difference by $\pi/2$ turns the cosine function in equation (4.112) into sine. Substituting for ϕ gives

$$I = \frac{I_0}{4} \left[1 + \sin \left(\sigma_x \frac{2\pi}{\lambda} C_\sigma nz \right) \right]. \tag{4.113}$$

Since ϕ is small this has two advantages. Firstly, near zero the sine function changes faster than cosine thus making the system more sensitive. Secondly, for small values the sine value of the phase and the phase expressed in radians are virtually the same. Equation (4.113) becomes

$$I = \frac{I_0}{4} \left(1 + \sigma_x \frac{2\pi}{\lambda} C_\sigma nz \right). \tag{4.114}$$

This means that σ_x becomes a linear function of the sensor output, which is preferable. Re-arranging equation (4.114) gives

$$\sigma_x = \frac{2\lambda I}{\pi C_\sigma nz I_0} - \frac{\lambda}{2\pi C_\sigma nz}. \tag{4.115}$$

The development of photoelastic optical fibre sensors began a generation ago. Spillman (1982) for example reported an application as a pressure sensor. Hydrophones and accelerometers were also developed in various research establishments. The US Naval Research Laboratory was interested in such devices in the early eighties. However, few systems came to commercial fruition and interest has somewhat declined currently.

4.5   Distributed sensing

Measuring some variable over an extensive physical range is an attractive concept to a number of areas of science and engineering. It is not a novel development. Electrical engineers have for several decades been using short voltage pulses to detect breaks in underground cables. A reflection occurs at the break and timing the return of the pulse enables an estimate to be made of the distance to the break. Commercial instruments are readily available for this. However, performance can be upset by the variation in the dielectric constant of the cable due to damage or ageing which affects the pulse speed. This procedure has been adopted by the telecommunications industry to locate breaks or poor joints in optical fibre lines. Radar and ultrasonic imaging are further examples of distributed sensing although not usually thought of as such. Optical fibre sensors are an attractive choice for distributed sensing not only because of the features listed in section 4.1 but also for the higher performance and, for the case of intrinsic sensing, the simplicity and compactness of the sensing line. A number of manufacturers now produce distributed optical fibre sensors. The earliest available and probably the most popular are for temperature measurement.

Optical fibre systems are classed as *distributed* when sensing takes place continuously along the fibre length. For systems where sensing occurs discretely at various locations the term *quasi-distributed* is used. The following applications have been identified as potential beneficiaries from distributed optical fibre sensors.

- Temperature measurement in process plant, power stations, marine systems and civil engineering structures such as long tunnels.
- Strain measurement particularly in civil engineering and aviation structures.
- Leak detection for pipelines, in basements and ducting, and for marine applications.
- Monitoring of composite materials and cements by embedded distributed sensors for either the forming (curing or setting) stages or in later use.
- Monitoring of electrical power engineering systems.
- Hydrophone arrays for submarine use.

In some applications, such as a fire alarms, a simple switching action is all that is required. Knowledge of how hot or where along the line the fire is may not be needed. A cryogenic leak detector has been marketed using the variation of refractive index with temperature to cause leakage of light from the fibre core into the cladding. The fibre is simply wrapped around the base of the cryogenic vessel or pipeline duct. When the temperature at any point along the fibre drops below $-60\,°C$ the alarm is triggered. Complex multiplexing circuitry is not needed—only continuous intensity monitoring of the light transmitted.

The problems facing designers of truly distributed optical fibre sensing systems are broadly twofold:

1. Identification of an appropriate sensing mechanism for the desired measurand.
2. Choosing a suitable multiplexing style so that data values returning from points along the line are separable.

4.5.1 Sensing mechanisms

A number of developments have been reported of sensing in distributed systems but not all have reached commercial exploitation. Several are described below. Fuller accounts can be found in Udd (1991) and Grattan and Meggitt (1995).

Rayleigh scattering

This form of scattering and its origins are described in section 4.2.2. Backscatter within the internal cone of acceptance is only a small fraction of the total scatter. Also, light travelling along the fibre in either direction is

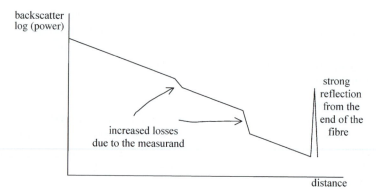

Figure 4.31. Attenuation of the back-scattered pulse.

attenuated exponentially. Consequently, the backscattered signal (figure 4.31) is weak and signal recovery techniques are needed to improve the signal-to-noise ratio.

The gradient discontinuities indicate increased loss at the two points illustrated. These are induced by the measurand and can be derived from a number of mechanisms such as

- the temperature dependence of backscattering,
- microbending,
- evanescent wave losses,
- temperature dependent absorption caused by a dopant in the glass,
- variation in birefringence in a monomode fibre due to the photoelastic effect or Kerr effect,
- rotation of polarization due to the Faraday effect.

Raman scattering

In 1928 Raman discovered that scattering by molecules not only produced Rayleigh scattering, where there is no change in wavelength, but also weaker, wavelength-shifted components. The effect was most easily observed in liquids, less so in solids and gases. These other wavelengths arise because of molecular vibrational and, in gases, rotational motions. An incoming photon causes the molecule to switch into an excited, *virtual* state. The word virtual is used here to mean an excited state that is not one of the specific energy states within the molecule. Rayleigh scattering occurs when the molecule returns to the ground state directly. The collision is elastic, the outgoing photon has the same energy as the incoming photon, and emission is isotropic. If, however, the molecule drops to an intermediate excited state, corresponding to a vibrational or rotational mode, the photon released has less energy. This is referred to as a Stokes photon. It has a longer wavelength (lower frequency) than the incoming photon. The probability of Raman scattering is orders of magnitude less likely than

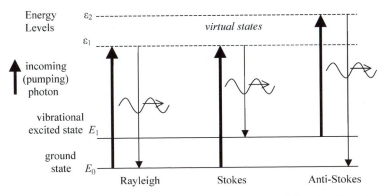

Figure 4.32. Energy level diagram for a molecule having a single (vibrational) excited state.

Rayleigh scattering and is an inelastic process. A further, even less likely, possibility is that the incoming photon strikes a molecule already in a vibrational or rotational excited state. When the molecule reverts to its ground state a photon of higher energy than the input photon is released. This is described as an anti-Stokes photon. Figure 4.32 illustrates these transitions in an energy diagram. The probability of the molecule being in its excited state before the collision is governed by the Maxwell–Boltzmann distribution with a value of $\exp\{-\Delta E/kT\}$, where ΔE is the energy interval $E_1 - E_0$.

The vibrational molecular frequency is modulated by the electric field of the incoming photon. The Raman frequencies of the emitted photons can be derived using the trigonometric identity given in chapter 3. If the molecular vibrational frequency is f_m and the photon frequency is f_0 then equation (3.14) gives the Stokes photon as having a frequency $f_0 - f_m$ and the anti-Stokes photon as $f_0 + f_m$. Figure 4.33 illustrates this as a wavelength diagram, more commonly used than frequency by optical engineers.

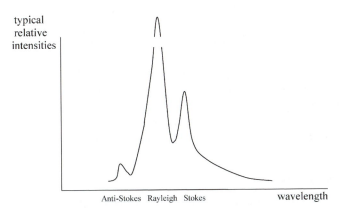

Figure 4.33. Representation of a scattering spectrum.

The shift in wavelength is sufficient in germanium-doped, silica fibres for high-quality, narrow-band filters to separate the light into its component parts.

It is found that the emitted power from the anti-Stokes band is the most sensitive to temperature at about 0.8% K^{-1} at 300 K. The usual practice is to measure the levels of Stokes and anti-Stokes backscatter using time-averaging techniques on the very weak signals received. The ratio is then determined as a referencing procedure to enable temperature measurement. Commercial devices are available giving a reasonably linear response in the range 200–400 K. Temperature resolution is about ±1 °C with a spatial resolution of a few metres in 1 km.

Brillouin scattering

Brillouin scattering is a similar phenomenon to Raman scattering except that the scattering is from groups of molecules in motion rather than individuals. The groups are moving in bulk because of acoustic vibrations in the material. The energy transitions are significantly smaller than those in the Raman phenomena, making spectroscopic resolution of the resulting wavelengths very difficult. Some sensor developments have occurred but commercialization of Brillouin-based systems looks less promising than for the other forms of scattering.

Polarization

In section 4.4.2 some of the phenomena affecting the state of polarization of light within fibres were described. In principle, any such phenomena could be useful in producing a distributed sensor. More sophisticated optical components and types of fibre are needed than a simple intensity-based system.

Quasi-distributed sensing

In principle any of the extrinsic or intrinsic sensing techniques described in previous sections could be used with an optical fibre line. The disadvantages to this approach are the possible bulk, fragility and costs of having an extensive bespoke system. Also, any multiple splicing or jointing will reduce the light intensity being propagated. The advantages are that, for a bespoke system, sensing for different measurands is possible as well as having different ranges or sensitivities at different locations. Also, the spatial resolution can be high, only limited by the multiplexing system. Having only a few discrete sensors means that the bandwidth of the fibre is available for precision—quality rather than quantity.

4.5.2 Multiplexing

Signal processing techniques in the well-established field of radar have been adapted for use in distributed optical fibre sensing. As with radar, the returned signal can be very weak and may be comparable with noise in the system. Use of a variable gain amplifier may be appropriate. The amplifier gain is made to be time-dependent so as to correct for attenuation loss with distance.

OTDR

Optical, time-domain reflectometry in its simplest form is a pulse-echo technique. A light pulse that is backscattered will return to the input end after a time delay proportional to the distance of the scattering region from the input. Effectively, figure 4.31 also illustrates the returned power versus time from an input pulse. For good spatial resolution short interval input pulses are required but this limits the pulse power which means a poor signal-to-noise ratio for the returned pulse. Averaging techniques, such as described in section 6.4.2, can be used to improve the signal-to-noise ratio. A strong reflection can be produced from the end of the fibre if this is suitably polished and coated, which could be useful in synchronizing the trigger for the averaging process.

A more sophisticated signal processing technique uses a random stream of binary pulses known as a *pseudo-random binary sequence* or PRBS. Such sequences are easily generated using various configurations of a shift register, an exclusive–OR gate and a clock. An example is shown in figure 4.34. The sequence is called pseudo-random because it is not truly random but repeats after a time dependent on the clock speed and the number of bits in the shift register. However, if the shift register is large enough the sequence will not repeat for years!

The maximum possible sequence length before repetition is $2^p - 1$ but to achieve this the correct value of n has to be chosen for a given p. Random code generation has received significant theoretical attention because of its usefulness in a number of signal processing applications and in analysing negative feedback systems. Commercial PRBS generators are available and are also often incorporated in signal analysers.

In the optical fibre distributed-sensor the PRBS is used to determine a time-correlation between the outgoing and returning signals. In chapter 6 the formal mathematical relationships for correlation are described. The

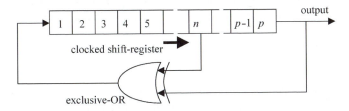

Figure 4.34. A simple PRBS generator.

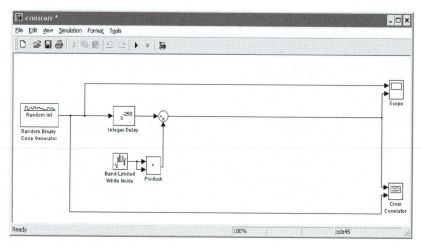

Figure 4.35. Cross-correlation simulation.

correlation algorithm in section 6.3.4 adapted for cross-correlation is

$$R_{xy}(r\Delta t) = \frac{1}{N}\sum_{j=1}^{N} x_j y_{j+r}. \qquad (6.44)$$

The outgoing pulse sequence x is stored in memory and the returned pulse sequence y is amplified and stored. Figure 4.35 shows a model of the system using the Matlab® toolbox Simulink. The integer delay between signals represents the transit time. Noise has been added, as seen in the scope trace of figure 4.36, to simulate the real fibre.

According to equation (6.44), the stored outgoing and returned sequences are multiplied together point-by-point to give a set of values which are then added together to give a particular value of R_{xy}. A time-shift of one clock pulse Δt is then made between the two stored pulse sequences and the multiplication–addition process repeated. A correlation is said to occur when one of the calculated values of $R_{xy}(r\Delta t)$ is a maximum. Knowing the number of clock shifts r needed to produce the correlation gives the time delay of the returned signal. Figure 4.37 illustrates the type of response.

A delay of 250 µs was inserted between signals. A correlation can be seen at the point 1.25 ms in the diagram. (The length of the time window is set by the correlator parameters whereas the start point is set by the simulation run-time.)

Polarization, OTDR

As mentioned in section 4.5.1, phenomena that cause changes to the state of polarization of light propagating along the fibre can be used for distributed sensing. Using pulses of light, as with time-domain reflectometry, the local

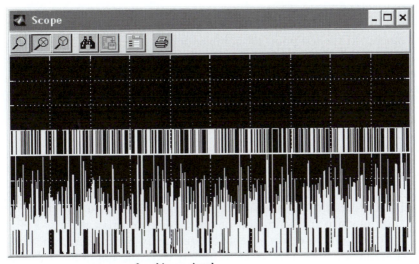

upper: random binary signal
lower: time-delayed, random binary signal + noise

Figure 4.36 Scope traces.

value of a measurand can be determined. The state of polarization is measured rather than the intensity. The Faraday and Kerr effects could facilitate the construction of distributed current and voltage sensors.

A more complex approach was reported in Dakin and Culshaw (1997). This strategy was devised to overcome the main disadvantage to using backscatter, namely the low intensity of the backscatter compared with the forward travelling wave (10^{-5}). Here the optical Kerr effect is used as part of the interrogation strategy. A continuous light wave is sent along the fibre in counter-propagation to a high-power, pulsed, light wave. The operational scheme is that the continuous wave has two modes of propagation which may be polarization states or waveguide modes, or even dual paths

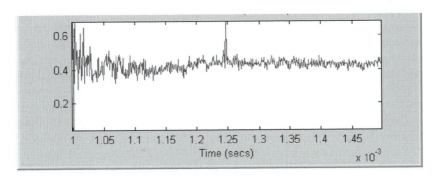

Figure 4.37. Output of the cross-correlator.

in two closely fabricated cores. The measurand must cause coupling to occur between modes. The optical Kerr effect also causes coupling of light between the modes. At the test point the combination of the effect of the measurand and the strong electric field from the pulse cause coupling nonlinearly. In this way the measurand and spatial information can be determined by intensity analysis of the two continuous output modes as a function of time. The system requires sophisticated optical and signal processing components. Variations of the strategy are still under development.

OFDR

Optical, frequency-domain reflectometry usually employs some form of *chirp*. A chirp is a varying frequency sinusoid. They have long been used by signal processing engineers in various applications including radar. The variation in chirp frequency can be any desired function such as a linear sweep, logarithmic sweep or a stepped sequence. In optical fibre applications a continuous-working, laser diode is usually used, the amplitude of which is modulated by the chirp. Because the outgoing pulse is frequency swept, the modulation frequency of the return pulse will be low for the nearer regions of the fibre and higher for the more distant. If a stepped chirp is used a Fourier transform of the detector output signal will deliver amplitude and discrete distance (frequency) information.

4.6   Referencing techniques

The majority of optical fibre sensors use the variation of light intensity to determine the value of the measurand. Consequently, for accurate values of the measurand precise measurement of the light intensity is required. However, the value of the output signal from the detector can vary for reasons other than the effect of the measurand. In order of significance these are as follows.

1. The emitted intensity of the source varying due to temperature or ageing. LEDs have lifetimes of over ten years but intensity variation with temperature is significant, typically $-0.7\%\,^{\circ}\mathrm{C}^{-1}$. Halogen lamps have a total operating time of about one year if used continuously, but this is reduced if the lamp is used intermittently or if subject to mechanical shock. Ageing of the filament was described in section 3.4.3. Also, a 10% change in supply voltage will cause a 4% change in the so-called *colour temperature*, i.e. the spectral distribution of figure 3.13, and a corresponding reduction in emission at the spectral peak of about 3%. This would not only have an effect on intensity measurement but would also change the relative intensities of the different colours present making ratiometric colour referencing unreliable. Finally, attenuation in

the fibre is dependent on the wavelength used, and is approximately $0.015\,\mathrm{dB/nm}^{-1}$ (figure 4.5). A change in wavelength of $1.0\,\mathrm{nm}$ would cause a 0.2% change in intensity received.

2. The components in the optical paths being affected by the environment. Introducing high curvature on fibres during installation can affect the modal distributions of the light and light leakage. Attenuation in connectors can vary by $\pm0.2\,\mathrm{dB}$. This could cause a variation of up to 5% of the full-scale deflection.

3. The sensitivity of the detector varying due to temperature fluctuation or component tolerance. PIN photo-diodes have lifetimes of over ten years. Their sensitivity to light varies at $+0.4\%\,^{\circ}\mathrm{C}^{-1}$ and is also wavelength dependent.

(The performance of sources and detectors was described in sections 3.5.3 and 3.5.4.)

Any combination of these sources of error can occur. Consequently, for high performance, some form of referencing is required if regular calibration is to be avoided. Some elaborate systems have been developed to minimize these effects. Generally, the more sophisticated the system the more bulky and expensive it becomes. Some of the more common techniques are outlined below. A list of references is given in Grattan and Meggitt (1995). Systems generally involve some form of ratiometric measurement for referencing.

Double detection

Two detectors can be used to monitor the behaviour of the source as illustrated in the simple system of figure 4.38.

The received intensity at the detectors can be expressed as

$$D_1 = \alpha MS, \qquad D_2 = \beta S \tag{4.116}$$

where α and β are calibration constants involving the optical lines. Eliminating S gives

$$M = \frac{\beta D_1}{\alpha D_2}. \tag{4.117}$$

This method relies on α, β and the sensitivity of the detectors being fixed. To reduce the effects of having different optical paths and detectors Y-connectors

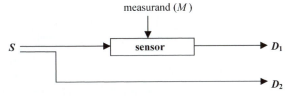

Figure 4.38. Referencing to compensate for source drift.

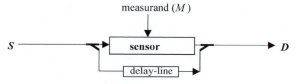

Figure 4.39. Multiplexing using a delay-line.

can be used. However, the use of Y-connectors can introduce further errors, add to the cost and may require the use of some form of multiplexing as described below.

Delay line

One simple multiplexing technique involves an optical delay-line as shown in figure 4.39. The source emits a short pulse that arrives as two separate pulses at the detector. High-speed signal conditioning circuitry is needed to measure the amplitude of each pulse. This technique is, in-effect, time-division multiplexing.

Wavelength modulation

If the effect the measurand has on the intensity of the light received is wavelength dependent, this can be used as a form of multiplexing. In tele-communications engineering this would be referred to as frequency-division multiplexing. The ratio of the wavelengths detected provides normalization. A single, broad-band source can be used as illustrated in figure 4.40(a), or two separate light sources as shown in figure 4.40(b).

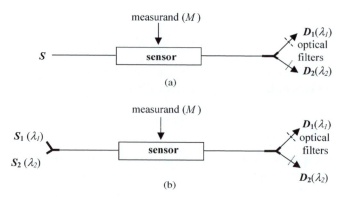

Figure 4.40. (a) Broad-band source, dual-wavelength referencing system. (b) Dual-source, dual-wavelength referencing system.

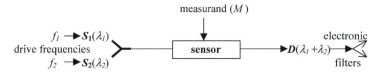

Figure 4.41. Dual wavelength referencing system using electronic filters.

Amplitude modulation

When using two sources an alternative approach avoiding the use of optical filters is to drive the sources at different frequencies (figure 4.41). Electronic filters are then used to separate the signals after the detector.

The optical bridge

A more optically sophisticated approach is the so-called optical bridge (figure 4.39). This takes its name from the electrical bridge circuit and involves two sources and two detectors. If the two sources are pulsed sequentially then

$$D_1 = \alpha M S_1, \qquad D_2 = \chi S_1 \tag{4.118}$$

and, a short time later,

$$D_1 = \beta S_2, \qquad D_2 = \delta S_2 \tag{4.119}$$

where α, β, χ and δ are the calibration constants of the optical paths. Combining these equations to eliminate S_1, S_2, D_1 and D_2

$$\frac{D_1}{D_2} = \frac{\alpha M}{\chi}, \qquad \frac{D_1}{D_2} = \frac{\beta}{\delta} \tag{4.120}$$

giving

$$M = \frac{\beta \chi}{\alpha \delta} \tag{4.121}$$

so that M only depends on the stability of the optical paths.

Figure 4.42 shows sources of different wavelength. With the TDM approach described above, identical sources and detectors would be appropriate. Also, of course, frequency modulation of two identical sources could be used with electrical filtering of the signals from two identical detectors.

Finally, it should be noted that the presence of ambient light might also contribute to the signal(s). This may be easily overcome by having the output of the source amplitude-modulated. Using a frequency of the order of 1 kHz will enable sunshine, lamplight at the power supply frequency and its harmonics to be removed. The desired signal is recovered by using a narrow-band,

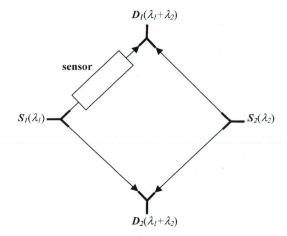

Figure 4.42. The optical bridge referencing system.

electronic filter in the detector signal processing circuitry set at the modulation frequency.

References

Bleaney B I and Bleaney B 1965 *Electricity and Magnetism* 2nd edition (Oxford University Press)

Dagenais D M, Bucholtz F and Koo K P 1988 'Elimination of residual signals and reduction of noise in a low-frequency magnetic fibre sensor' *Appl. Phys. Lett.* **53** 1417–1476

Dakin J and Culshaw B 1988 *Opical Fiber Sensors: Principles and Components* vol 1 (Boston: Artech House)

Dakin J and Culshaw B 1997 *Opical Fiber Sensors: Applications, Analysis and Future Trends* vol 4 (Boston: Artech House)

Dotsch H, Martens G and Meyer W 1983 *Fibre Optic Components for Industrial Control*, IEE Conf. pub. 221, *Optical Fibre Sensors* (London: IEE) pp 67–71

Frank W E 1966 'Detection and measurement devices having a small, flexible fiber transmission line' US Patent 3,273,447 September

Göpel W, Hesses J and Zemel J N 1992 *Sensors: a Comprehensive Survey* vol 6 *Optical Sensors* (New York: VCH

Grattan K T V and B T Meggitt (eds) 1995 *Optical Fiber Sensor Technology* (Chapman and Hall)

Gupta G C 2002 *Optical Fiber Communication Conf., Anaheim, CA*, paper TuA5

Jones B E 1981 'Simple optical sensors for the process industries using incoherent light' *Symp. Optical Sensors and Optical Techniques in Instrumentation, London, UK* (London: IMC) p 15

Jones B E 1985 'Optical fibre sensors and systems for industry' *J. Phys. E: Sci. Instrum.* **18** 770–782

Jones B E and Philp G S 1983 'A fibre optic pressure sensor using reflective techniques' *Proc. Int. Conf. Optical Techniques in Process Control, The Hague, Netherlands* (Cranfield: BHRA Fluid Engineering) pp 11–25

Keck D B, Maurer R D and Schultz P C 1973 *Appl. Phys. Lett.* **22** 307

Kirkbright G F, Narayanaswamy R and Welti N A 1984 'Fibre-optic pH probe based on the use of an immobilised colorimetric reagent' *Analyst* **109** 1025

Kissinger C D 1967 'Fibre optic proximity probe' US Patent 3,327,584, June

Lakowicz J R 1983 *Principles of Fluorescence Spectroscopy* (New York: Plenum Press)

Land H E 1951 'Some aspects of the development of sheet polarizers' *J. Opt. Soc. Am.* **41**(12) 957

Livingston J D 1982 'Magnetomechanical properties of amorphous metals' *Phys. Status Solidi* **70** 591–596

Longhurst R S 1970 *Geometrical and Physical Optics* 2nd edition (Longmans)

Nishihara H, Haruna M and Suhara T 1989 *Optical Integrated Circuits* (New York: McGraw-Hill)

Nishizawa K, Sudo E, Yoshida M and Yamasaki T 1986 'High sensitivity waveguide type hydrogen sensor' *4th Int. Conf. Optical Fibre Sensors, Tokyo* p 131

Ohashi M, Shiraki K and Tajima K 1992 'Optical loss property of silica-based single mode fibers' *IEEE J. Lightwave Technol.* **10**(5) 539–543

Pérot A and Fabry C H 1898 'Sur un voltmètre electrostatique interférentiel pour etalonnage' *J. de Phys.* **7** 650–659

Post E J 1967 'Sagnac effect' *Rev. Mod. Phys.* **39** 475

Rogers A J 1979 'Optical measurement of current and voltage on power systems' *IEE J. Electronics and Power Appl.* **2**(4) 120–124

Scheggi A M and Baldini F 1993 'Chemical sensing with optical fibres' *Int. J. Optoelectronics* **8**(2) 133–156

Senior J M 1985 *Optical Fibre Communications, Principles and Practice* (Prentice-Hall)

Spillman W B jnr 1982 'Multimode fibre-optic pressure sensor based on the photoelastic effect' *Opt. Lett.* **7** 388

Spillman W B jnr and McMahon D H 1980 'Frustrated-total-internal reflection multimode fibre-optic hydrophone' *Appl. Opt.* **19** 113–117

Udd E (ed) 1991 *Fiber Optic Sensors, An Introduction for Engineers and Scientists* (Wiley-Interscience)

Yamada J I, Saruwatari M, Asatani K, Tsuchiya H, Kawana A, Sugiyama K and Kimura T 1980 'High speed optical pulse transmission at 1.29 µm wavelength using low-loss single-mode fibers' *IEEE J. Quantum Electronics* **QE-14** 791

Exercises

1. Compare the performance of metallic telemetry lines with those of optical fibre. List the reasons why optical fibre sensors are attractive.

2. Explain the significance of the numerical aperture when dealing with optical fibres and derive the formula.

3. (a) Attenuation in optical fibres can be caused by absorption and scattering. Describe the nature of these phenomena and include a sketch graph.

   (b) A test facility is to be built to investigate wavelength related attenuation in optical fibres.

  (i) Draw a diagram of an appropriate arrangement suggesting a suitable light source and detector.

  (ii) Draw a signal conditioning circuit for your detector and explain its operation. (Hint: see chapter 3.)

  (iii) How might referencing be achieved to overcome drift in the system performance?

4. Explain how black-body radiation can be used to measure temperature by means of an optical fibre.

 (a) Draw a diagram naming the materials and optical components of such a device.

 (b) Drift in the detector and optical couplings can be accounted for by using ratiometric measurement. Using the Planck formula given in section 3.5.3, determine the intensity ratio for light of wavelengths 550 nm and 450 nm at a temperature of 700 °C and then at 725 °C. (You may assume that the filters used are non-attenuating and have the same very narrow bandwidth.)

 (c) How might the problem of the sensor probe tip not being a perfect black-body be accounted for?

5. Two optical fibres are placed side-by-side to form an intensity-based, reflectance-mode, displacement sensor. Their core radii are a_1 and a_2 and the separation of their centres is D.

 (a) Explain, with the aid of a diagram, how movement of the reflector can produce a near-linear variation in intensity with displacement. Derive two expressions I_{min} and I_{max} for the limits of this near-linear, displacement range.

 (b) How might such a device be used to measure temperature? Draw a sketch.

6. An optical-fibre chemical sensor is to be developed based on the absorption principle.

 (a) State the Lambert–Beer absorption law.

 (b) With the aid of a sketch describe a suitable test-cell structure.

 (c) By judicious choice of the signal conditioning circuit a photodiode detector could be made to output a linear response with concentration. Explain this quoting any appropriate formulae. (Hint: see photodiode section in chapter 3.)

7. Design an evanescent, front-end displacement sensor after the style of Figure 4.19. Determine an appropriate oblique angle for the front face using typical values for the refractive indices etc. Determine the sensitivity for your design as the intensity-displacement gradient, i.e. dI/dx.

8. The Faraday magneto-optic effect and the Kerr effect can be used to measure electrical current and field. Sketch an arrangement whereby

both of these phenomena could be implemented in one optical fibre sensor so that, in principle, an electrical power sensor could be built for use on a high-voltage overhead power line. What would be the major difficulty to be overcome?

Chapter 5

Intelligent sensing

5.1 Introduction

The development of the microprocessor and the subsequent IBM personal computer (PC) led to a growth in the processing of the data from sensors during the 1980s. Many data acquisition boards or *cards* have become available for plugging into the IBM-type bus with sophisticated support software for data logging, processing, analysis and display. Such activity has been described as intelligent instrumentation. More recently sensors have become available having a microprocessor within the sensor head enabling processing to be carried out *in situ*. These devices have come to be termed *smart* sensors and are outlined in section 5.5. Artificial intelligence (AI) software is also finding application in instrumentation systems for interpreting the output of one or more sensors. Some of the developments of AI are briefly described in section 5.6.

Before considering the application of intelligence to sensing it is important to appreciate the restrictions placed on the empirical data caused by the interfacing to a digital system. The output of a sensor will generally be an analogue, smoothly-varying function with time. However, to incorporate this signal into a digital system it has to be limited in two ways:

- It must be time-sampled. This is sometimes called discrete time-sampling. During the processing of the sampled value of the measurand the system is off-line and so empirical data will be lost.
- Because a digital system has a prescribed number of bits, resolution of the empirical data will be reduced. The continuous span of the input measurand becomes discrete in value.

Time-sampling in PC-based systems may be deliberately limited to moderate rates so that significant processing and outputting can be carried out between the taking of the samples. Such a system would be described as operating in *real time* because there is no need for archiving along the data acquisition pipeline. Provided that the run-time of the algorithm doing the processing is not too involved, sample rates of up to 10^5 per second are possible. However, graphical interfacing to the monitor is slow and when visual presentations of the data are required this will severely reduce the acquisition speed if the system is to remain real-time.

The sampled data must be changed to digital format using an analogue-to-digital converter (ADC). In a simple system this will be under the control of the PC. However, in systems working at speeds beyond the limit of the PC, a separate board with its own microprocessor is used. The ADC and any extra memory are controlled locally on the board. The board may also process the data. Specialist digital signal processing (dsp) microprocessors are commonly used with algorithms for digital filtering or Fourier analysis etc. (see section 5.4). In this way large volumes of data, sampled at high speed, can be stored and processed. The processed information can then be sent to the PC at a slower rate for further analysis and display. Such a system would be described as real-time if the information was flowing through each successive section of the system at the same rate.

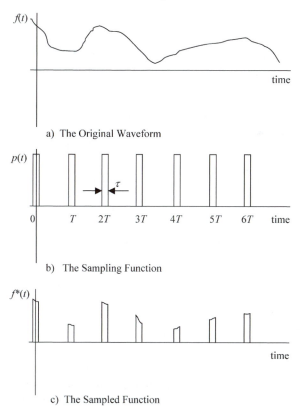

a) The Original Waveform

b) The Sampling Function

c) The Sampled Function

Figure 5.1. The time-sampling of a continuous waveform.

The problems related to discontinuous and discrete sampling are described in the following sections.

5.2   Discontinuous-time sampling

When a waveform $f(t)$ is time-sampled at some rate a new signal is produced which is, in effect, an amplitude modulated version of the original:

$$f^* = p(t)f(t). \tag{5.1}$$

The function $p(t)$ is a train of narrow unit pulses at a frequency $\omega_s(t)$. This process has the effect of producing *sidebands* in the frequency spectrum as illustrated in figure 5.2.

The sidebands occur centred at multiples of the sampling frequency as shown. To avoid the sidebands overlapping with the original, the sampling frequency must be chosen with reference to the maximum frequency ω_{max} within the function $f(t)$. From the diagram it can be seen that for the centre spectrum not to touch one of the sidebands,

$$\omega_s \geq 2\omega_{max}. \tag{5.2}$$

This result was discovered independently by Shannon and Nyquist. It can be formally verified as shown below. Using complex notation for the Fourier series, $p(t)$ can be written as

$$p(t) = \sum_{k=-\infty}^{\infty} c_k \exp jk\omega_s t \tag{5.3}$$

where

$$c_k = \frac{\tau}{T} \frac{\sin(k\pi\tau/T)}{k\pi\tau/T} \tag{5.4}$$

or

$$c_k = \frac{\tau}{T} \operatorname{sinc}\left(\frac{k\pi\tau}{T}\right). \tag{5.5}$$

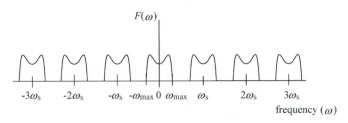

Figure 5.2. The frequency spectrum of an ideally sampled function.

T is the sampling period and τ is the width of the sampling pulse. The time-sampled function of $f(t)$ is

$$f_s(t) = f(t)p(t) \tag{5.6}$$

$$= f(t) \sum_{k=-\infty}^{\infty} c_k \exp jk\omega_s t. \tag{5.7}$$

The Fourier transform of $f_s(t)$ is

$$\Im[f_s(t)] = \int_{-\infty}^{\infty} \sum_{k=-\infty}^{\infty} c_k \exp jk\omega_s t f(t) \exp -j\omega t \, dt \tag{5.8}$$

$$= \sum_{k=-\infty}^{\infty} c_k \int_{-\infty}^{\infty} f(t) \exp -j(\omega - k\omega_s)t \, dt \tag{5.9}$$

$$\equiv \sum_{k=-\infty}^{\infty} c_k \Im[f(\omega - k\omega_s)]. \tag{5.10}$$

The interpretation of equation (5.10) is that the Fourier transform of $f_s(t)$ is the Fourier transform of $f(t)$ but is repeated at multiples of ω_s and that these multiples are attenuated by differing amounts according to the function c_k. This confirms the relationship stated in equation (5.1) and in addition shows that for a sampling function that is not made from infinitely narrow pulses, the sidebands are attenuated by a sinc function.

If a system is designed for input frequencies up to ω_{max}, having $\omega_s = 2\omega_{max}$, and the signal increases to $\omega_{max} + \delta\omega$, then a lower frequency is produced at the system output of $\omega_{max} - \delta\omega$. This is known as *aliasing*. Aliasing is the production of low frequencies in the output by the sampling system itself. In systems that have a fixed sampling rate an anti-aliasing filter may be needed to limit the input frequencies to $< \frac{1}{2}\omega_s$. Because realisable filters do not have sharp cut-off frequencies it is necessary to allow a greater frequency interval between the spectrum and the sidebands. High fidelity, real systems may have sampling frequencies of $6\omega_{max}$.

5.3 Discrete sampling of the measurand

Any digital word is binary weighted with the most significant bit (MSB) having the highest weighting and the least significant bit (LSB) the lowest weighting. It follows therefore that the best resolution that can be achieved when converting from an analogue value to a digital word is $\leq \frac{1}{2}$LSB. This is sometimes referred to as the quantization error and can be up to $\pm\frac{1}{2}$LSB. Figure 5.3 illustrates the effect.

Consequently, when designing a data acquisition system the required resolution is used to determine the number of bits needed in the digital

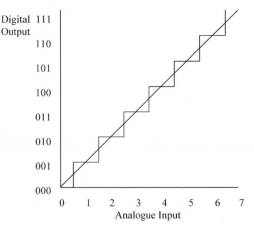

Figure 5.3. An ADC transfer curve.

word. For example, an 8-bit ADC able to accept analogue inputs within the range 0–10 V will have a resolution of

$$\pm \frac{1}{2} \times \frac{10}{255} \text{ V} \quad \text{or} \quad \pm 0.01961 \text{ V}.$$

As well as the $\frac{1}{2}$LSB quantization occurring across the range it also creates a zero error of $\frac{1}{2}$LSB as shown in figure 5.3.

5.3.1 Digital-to-analogue conversion

Digital-to-analogue converters (DACs) are described here before ADCs since some ADCs use a DAC as part of their structure. A common use of the DAC is for the conversion of the digital output of a computer system into an analogue value. Another use in instrumentation is with digital sensors when analogue processing/transmission of the output is required. Examples of digital sensors are the Gray code disc, for encoding the position of a rotating shaft and the multi-probe conductivity sensor for liquid level measurement.

A digital word can be converted into an analogue current or voltage by a variety of switching circuits. The most commonly used circuit is the R–$2R$ ladder network. The circuit shown in figure 5.4 has an op-amp as an output buffer summing the switched currents to provide a voltage output. Without the op-amp the circuit will act as a current source.

The R–$2R$ configuration ensures that the current is halved at each junction. Currents are switched between ground and the virtual ground of the op-amp. Thus, provided the reference voltage source V_{ref} and the resistors remain constant, the fractional weightings of the currents will be correct and their summation in the op-amp will provide the correct analogue

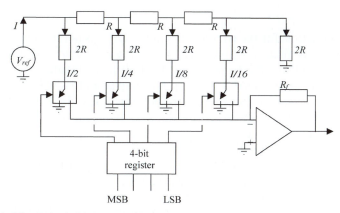

Figure 5.4. The *R–2R* ladder network.

output. The output will, of course, be stepped rather than smoothly varying. If this is undesirable then placing a low pass filter on the output will remove the step effect.

There are two main types of error that may occur with DACs. These can best be appreciated using the graph in figure 5.5. If the staircase curve were not to be properly aligned with the straight line (perhaps due to a drift in the internal voltage source V_{ref}) then there would be a calibration error across the range. The error could be up to 1 LSB at full scale. The other error is to do with nonlinearity (this might be due to poor tolerances or drift of the resistor values). Errors occur across the range and could be $\frac{1}{2}$ LSB or greater.

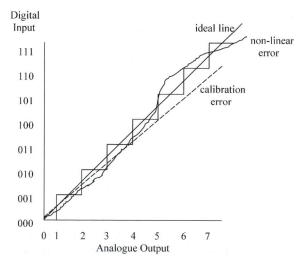

Figure 5.5. Types of error with DACs.

Settling time

The switching time for the components in the DAC is not infinitely fast. The settling time is that time required, following an input change, for the analogue output to settle to within $\pm\frac{1}{2}$LSB of the final value.

In current output DACs the output is converted to a voltage by simply connecting an external load resistor to the current line. In this type of DAC output settling-time is governed by the time constant associated with the external load resistance, the DAC output capacitance and any stray capacitance in parallel with the external load resistor.

In voltage output DACs the op-amp buffer gives the advantage of low output impedance but at the cost of reduced speed. The slew rate of the op-amp (the maximum rate of change of output voltage) has a major influence on the system settling time. Consequently, the LSB will have the shortest settling time and the MSB will have the longest. Also, the external circuitry has a more pronounced effect on the settling time than that for the current output DAC and may contribute significantly to it.

Glitches

Glitches in digital systems are transient voltage spikes, positive or negative going, occurring because of shortcomings in the switching process. The time span of a glitch is typically of the order of a nanosecond. Because of this, and the somewhat random nature of their occurrence, their detection can be extremely difficult. This makes faultfinding in a working system, using a conventional oscilloscope or analyser, an unenviable task.

Glitches can occur at the analogue output of a DAC when a transition is occurring in the MSBs. Figure 5.6 illustrates the possible outcome with a 3-bit DAC.

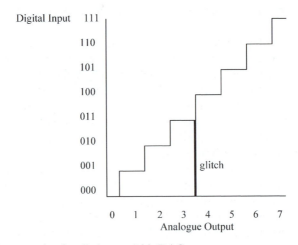

Figure 5.6. An example of a glitch on a 3-bit DAC.

Table 5.1. The Gray code.

Deanery number	Standard binary code value	Gray code value
0	0000	0000
1	0001	0001
2	0010	0011
3	0011	0010
4	0100	0110
5	0101	0111
6	0110	0101
7	0111	0100
8	1000	1100
9	1001	1101
10	1010	1111
11	1011	1110
12	1100	1010
13	1101	1011
14	1110	1001
15	1111	1000

On being incremented by 1 LSB the input value 011 should change to 100,

011

(000 glitch)

100

but a large output glitch may occur of 000 if the 1s become 0s before the 0 becomes a 1.

Binary codes have been developed to minimize the transitional error that can occur. The Gray code, mentioned above in the context of a rotational sensor, is one such code. At each transition the bit is changed that would give the least error if mis-read. Some values are shown in table 5.1.

Glitches are less likely to occur in fast-switching systems. If necessary, glitches may be removed using a zero-order-hold (ZOH) circuit. The expression *zero order* is used to denote that the device does not modify the value it is holding. Other hold devices attempt to correct for changes that are occurring while off-line. They apply a correction related to a first or higher order polynomial. Figure 5.7 illustrates the schematic diagram of a ZOH. The DAC is connected to the capacitor until just before a conversion is ready. The switch is then opened to isolate the DAC. During this time the capacitor provides the output value. After a suitable settling time the switch is closed again and the capacitor value and output up-dated.

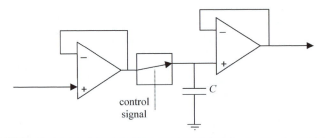

Figure 5.7. The zero-order-hold device.

This process introduces an extra delay and so slows the overall rate of conversion.

5.3.2 Analogue-to-digital conversion

There are a large number of ADCs on the market using a variety of techniques. Prices vary from a few pounds to over £100. The highest prices are paid for ADCs with high resolution and high speed. Prices continue to fall and this is particularly the case for those enjoying mass application. ADCs used in video systems are an example of a higher performance device being reasonably cheap. The following sections illustrate three broad groupings of ADCs according to their operating technique, performance and relative cost.

1. The parallel or flash conversion ADC

These devices are the fastest and most expensive ADCs. If high resolution is also required cost is high. The technique involves a reference voltage source which is connected across a potential divider chain to provide a set of reference voltages. The analogue input is simultaneously compared with all of these values using a set of comparators. Comparator circuits operate at high speed giving a fast result, typically <30 ns. The comparator outputs are encoded into an n-bit digital word. For an 8-bit ADC the converter would require 255 comparators. The schematic circuit for a 3-bit device is shown in figure 5.8.

For very high rate data acquisition systems, where even this device is not fast enough, the signal may be multiplexed to several such ADCs so that the conversion time does not limit the acquisition rate.

2. The dual slope ADC

The dual slope conversion technique is a slow procedure. However, high-resolution devices can be purchased cheaply. (See figure 5.9.)

The integrator is initially connected to the analogue input for a fixed time period, t_1. The control unit then switches the integrator input to a reference voltage of opposite sign and the time taken t_2, to discharge the

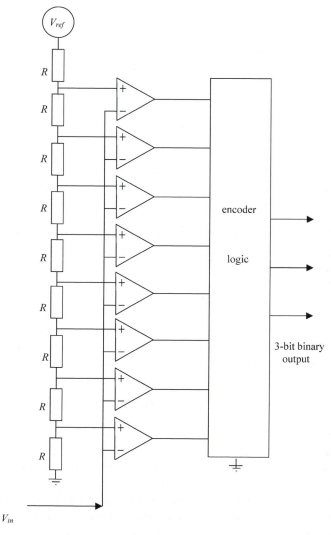

V_{in}

Figure 5.8. The parallel conversion ADC.

capacitor is determined. The timings (figure 5.10) are carried out using a clock, a logic unit and a counter. During the time t_1, the capacitor acquires the potential difference V, where

$$V = -t_1 \frac{V_{in}}{RC} \tag{5.11}$$

$$V = -\frac{N_0}{f_c} \frac{V_{in}}{RC} \tag{5.12}$$

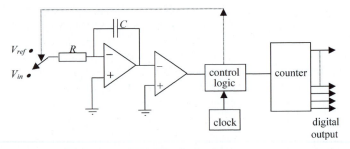

Figure 5.9. Simplified schematic for the dual slope ADC.

where f_c is the clock frequency and N_0 is the number of clock cycles that sets the charging time t_1.

During the integration of the reference source the capacitor must suffer the same change in potential but of opposite sign so as to become discharged. That is

$$V = t_2 \frac{V_{\text{ref}}}{RC} \tag{5.13}$$

$$V = \frac{n}{f_c} \frac{V_{\text{ref}}}{RC} \tag{5.14}$$

where n is the variable number of clock cycles required to discharge the capacitor. Setting equation (5.12) equal to (5.14) for complete discharge,

$$\frac{N_0}{f_c} \frac{V_{\text{in}}}{RC} = \frac{n}{f_c} \frac{V_{\text{ref}}}{RC} \tag{5.15}$$

or

$$n = N_0 \frac{V_{\text{in}}}{V_{\text{ref}}}. \tag{5.16}$$

This shows that the value of the count n is directly proportional to the analogue input V_{in}. High precision of the values of R and C and other components is not required. However, component stability is essential. Also, variation in switch resistance and the reference voltage source will

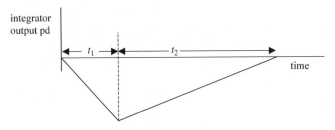

Figure 5.10. Dual slope ADC timing sequence.

significantly affect accuracy. The main disadvantage is that the conversion time $(t_1 + t_2)$ is long, typically ≥ 40 ms and is dependent on the input level.

This relatively simple technique provides high resolution chips at low cost. They are widely used in digital electronic instruments such as digital multimeters and phase meters. Because of this well-established application area these ADC chips often contain additional features such as auto-zeroing. Also, in applications where interference may be a problem, the value of t_1 can be set to an integer number of cycles of the interfering frequency. Integrating V_{in} over this period would tend to average the interference amplitude to zero.

3. Feedback converters

In terms of performance, feedback converters fall between the flash converters and the dual slope devices. They are the most popular ADCs, with a broad range of specifications available from many manufacturers. As the name implies a conversion is performed which is then fed back through a DAC and a comparison made with the original analogue input. The many types of feedback converter can be classified according to the sequence by which the output is generated.

3a. The successive approximation converter

This is one of the most popular of all of the types of ADC (figure 5.11). The operating procedure is sometimes called a *binary search*. The bits in the generated binary word are set high in sequence and at each change their value is compared with the input.

At the start of conversion the logic circuitry sets the MSB high and then waits for the decision from the comparator. If the digital output is larger than the analogue input the MSB is set to zero. If the digital output is less than the analogue input the MSB is retained at 1 and the process is repeated on the next lower MSB. Each new test effectively is evaluating half of the range of the previous one. Because of this the process is also referred to as

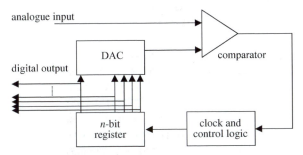

Figure 5.11. The successive approximation ADC.

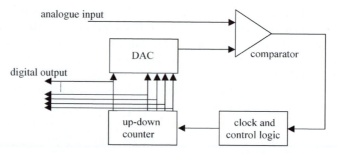

Figure 5.12. The tracking ADC.

an *interval-halving technique*. For an *n*-bit digital output there will be *n* comparisons made. This means that the conversion time for this type of feedback converter is fixed regardless of the behaviour of the analogue input.

3b. The tracking converter

In this ADC an up-down counter (figure 5.12) provides the digital output. A comparator again monitors the input and feedback signals and controls the up/down incrementing as appropriate.

    If the digital output signal is too high the counter will decrement its output by 1 LSB and, if too low, the counter will increment its output by 1 LSB. Usually, the comparator has thresholds so that the digital output does not change if the error is less than $\pm\frac{1}{2}$LSB. The time to generate the conversion is one clock cycle.

    As long as the input is not changing by more than ±1 LSB during a clock cycle the converter will correctly output the digital equivalent of the analogue input. An illustration is given in figure 5.13.

    A summary is given below in table 5.2 of the ADC types and their performance. There are a number of other types of ADC available. Some, like the dual slope converter, are based on charge balancing; examples are the switched capacitor and the *delta-sigma* systems. Others employ single-slope

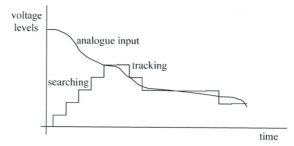

Figure 5.13. Performance of a tracking ADC.

Table 5.2. Comparison of some ADC types.

Analogue-to-digital converter	Cost	Speed	Comments
Parallel or flash	High	Very fast (30 ns)	Used in fast, high-performance systems. Up to 10-bit versions available, but 8-bit is the most common.
Dual slope	Low	Slow (40 ms)	Variable conversion time. Can be configured to reduce interference. Good accuracy, 10-bit to 18-bit versions are available.
Feedback converters	Moderate to high	Fast, 25 μs to 400 ns	Up to 16-bit resolution is available.

integration or voltage-to-frequency conversion. See Horowitz and Hill (1989) for descriptions.

Conversion speed and frequency response of ADCs

ADCs are designed to accept either bipolar or unipolar input, such as ± 5 V or 0–10 V and come with resolutions of 6, 8, 10, 12, 16 or 18 bits. If the full resolution of an ADC is to be utilized the signal input level should be close to the maximum input voltage that the ADC can accept. Also, to maintain resolution to 1 LSB the amplitude of the analogue input signal must not change by more than the equivalent of $\pm \frac{1}{2}$ LSB during the time in which the ADC performs the conversion. To determine the input frequency limit that can be offered to the device an analysis is performed using a sinusoidal input

$$V = V_0 \sin \omega t. \tag{5.17}$$

Differentiating to determine the slew rate,

$$\frac{dV}{dt} = \omega V_0 \cos \omega t. \tag{5.18}$$

The maximum slew rate is therefore

$$\frac{dV}{dt} = \omega V_0. \tag{5.19}$$

The implications of the criteria described above can be seen in the analyses given below. The ADCs are assumed to be unipolar. Consequently, because a sinusoid has a maximum amplitude swing of $2V_0$ for best performance,

$$2V_0 = V_{max} \tag{5.20}$$

where V_{max} is the maximum permitted input voltage.

Generally, ADCs such as the flash converter and the tracking converter can change their output at the rate of 1 LSB per clock cycle. But, as already stated the correct digital output will only be delivered as long as the rate of change of the analogue input does not exceed $\pm\frac{1}{2}$ LSB per clock cycle. Hence the maximum acceptable slew rate is given by

$$\frac{dV}{dt} = \frac{\frac{1}{2}\,\text{LSB}}{\text{clock period}}. \tag{5.21}$$

Now,

$$1\,\text{LSB} = \frac{V_{max}}{2^n} \tag{5.22}$$

2^n being the number of increments that V_{max} is divided into. Also, the clock rate f_c can be substituted for the clock period,

$$\text{clock period} = \frac{1}{f_c}. \tag{5.23}$$

So equation (5.21) becomes

$$\frac{dV}{dt} = \frac{\frac{1}{2}V_{max}}{2^n}f_c \tag{5.24}$$

$$\frac{dV}{dt} = \frac{V_{max}}{2^{n+1}}f_c. \tag{5.25}$$

Equating the two slew rates of equations (5.19) and (5.25) gives

$$\omega V_0 = \frac{V_{max}}{2^{n+1}}f_c. \tag{5.26}$$

Substituting for V_{max} using equation (5.20),

$$\omega V_0 = \frac{2V_0}{2^{n+1}}f_c. \tag{5.27}$$

This sets the maximum value of the input frequency ω_{max} as

$$\omega_{max} = \frac{f_c}{2^n}. \tag{5.28}$$

For an 8-bit resolution ADC with a clock rate of rate of $10^6\,\text{s}^{-1}$, equation (5.28) gives

$$f_{max} = 621.7\,\text{Hz}. \tag{5.29}$$

In this example the digital output is updated every 1 µs. But the analogue input frequency must not exceed the 621.7 Hz. Also, of course, no fast transients or steps must occur. This is in stark contrast to the frequency limit set by the Shannon–Nyquist sampling theorem described in section 5.2. The sampling theorem states that $f_{max} < 2f_s$, the term f_s being the sampling rate. In this example the sampling rate is the clock rate of $10^6 \, \text{s}^{-1}$ which sets the limit of f_{max} as 500 kHz.

A similar analysis can be performed for the successive approximation converter. As explained above an n-bit device would take n clock cycles to complete a conversion. As before, while the ADC is performing the conversion the analogue value must not change by more than the equivalent of $\pm\frac{1}{2}$LSB. Again, considering the slew rates as previously,

$$\frac{dV}{dt} = \frac{\frac{1}{2}\,\text{LSB}}{n \text{ clock cycles}} \tag{5.30}$$

so the maximum frequency reduces to

$$\omega_{max} = \frac{f_c}{n2^n}. \tag{5.31}$$

Using the same numerical values as the previous example, a resolution of 8-bits and clock period of 1 μs then

$$f_{max} = \frac{10^6}{8\pi 2^9} \quad \text{or} \quad 77.71 \,\text{Hz}. \tag{5.32}$$

The conversion time is 8 μs, so the sampling rate can only be up to 125 ksamples s^{-1}. For this sampling rate the Shannon–Nyquist theorem sets the maximum frequency input as 62.5 kHz.

Clearly, restricting the movement of the analogue input to $\pm\frac{1}{2}$LSB during conversion to provide good resolution is limiting the frequency performance of the ADCs. This restriction may be overcome by having a device before the ADC which samples the analogue signal and inputs this as a fixed value to the ADC for the length of time it takes to perform the conversion. This holding technique is described below.

The sample-and-hold device

The zero-order-hold device of figure 5.7 and the sample-and-hold device are functionally very similar. The need to use a sample-and-hold device will depend on the frequencies present in the analogue input signal and the resolution required. Sample-and-hold devices are not infinitely fast at sampling the signal but this *aperture time*, as it is known, can be typically as low as 0.3 ps. This effectively improves the acceptable input slew rate to

$$\frac{dV}{dt} = \frac{\pm\frac{1}{2}\,\text{LSB}}{\text{aperture time}}. \tag{5.33}$$

Equating this with the maximum slew rate of a sinusoid given in equation (5.19) gives

$$\omega V_0 = \frac{\frac{1}{2}(2^{-n}V_{max})}{\text{aperture time}}.$$

(5.34)

As before, if the full input range of the ADC is used $2V_0 = V_{max}$ giving

$$\omega_{max} = \frac{1}{2^n(\text{aperture time})}$$

(5.35)

which is similar to previous expressions for ω_{max} but the aperture time has replaced the clock period. Using the previous example of an 8-bit device and assuming an aperture time of 0.3 ps, equation (5.35) gives f_{max} as 2072 MHz! Clearly, a marked improvement has occurred in the maximum acceptable input frequency.

The maximum permissible input frequency is now only limited by the Shannon–Nyquist sampling theorem. All ADCs take some finite time τ to perform the conversion. During this time they are off-line and so this effectively sets the sampling rate. If a sample-and-hold device is used then the sampling time is slightly lengthened. The Shannon–Nyquist theorem gives the maximum acceptable input frequency as

$$f_{max} < \frac{1}{2(\tau + \text{aperture time})}.$$

(5.36)

To conclude, the use of a sample-and-hold device enables the theoretical limit set by the sampling theorem to be realized without deterioration in ADC resolution.

5.4   PC-based data acquisition boards

There are a large number of boards now commercially available and consequently it is worthwhile considering the criteria for choosing a particular product. Boards range in price from less than £100 to several thousand pounds. The large price range is due to the variation in speed, resolution and the sophistication of the data handling. A simple board would typically have an ADC of 12-bit resolution, with a conversion time of 1 ms and no data handling capability other than simply transferring information to the PC bus. As well as having a well-known brand name, a board at the top of the price range will have some memory and processing power of its own and perhaps be able to write directly to the hard drive. However, the most significant feature to increase cost is the sampling speed.

A high speed board might have an ADC of 16-bit resolution operating in <10 μs. Such ADCs alone can cost in excess of £100. There may be one such ADC on each channel since multiplexing of input channels through one ADC converter slows the sampling rate. For this reason the highest speed boards usually only have a few input channels. When reading manufacturers' specifications it is important to distinguish between the conversion rate of the ADC and the actual sampling rate of a particular input channel. Clearly, a board with a single ADC and 16 input channels will be sampling at least 16 times slower than the ADC. Some boards will allow the multiplexer to be turned off so that the one functioning channel has the full ADC speed.

To appreciate the way the PC microprocessor communicates with these boards a little knowledge of microprocessor system architecture and the handling of data input/output is needed. Broadly speaking a microprocessor uses two designated regions of memory, one for storing the program and the other for the data being processed. When the computer system boots up, the program is loaded to the microprocessor's electronic memory from a non-volatile memory (not lost at power down) such as a hard disk. The electronic memory is provided within the microprocessor itself and also externally on arrays of memory chips. Electronic memory is used because of its far higher access speed compared with that of memory media used for archiving. The microprocessor has an addressing system for access to these electronic memory locations. During the design of the microprocessor chip and its peripherals the addresses of the various locations of memory are designated. This is referred to as the memory map. Within the data memory mapped locations, several addresses will be input/output ports and not actual memory. When the program is running and data are collected from or sent to these particular locations the microprocessor is effectively communicating externally.

Software to transfer digital data via the PC ports can be written in a high level language such as Pascal or some version of C. This is relatively straightforward and when using commercially available data acquisition and processing boards is usually supplied with the board by the manufacturers. The example given below illustrates the use of an *inline* routine to access data presented to a PC input port from a data acquisition and processing board. The expression inline refers to a section of code written in assembly language within a program written in some high-level language. This involves the direct use of the instruction set for the microprocessor as devised by the chip designers. However, because every data movement and processing action has to be specified individually it is a tedious activity and time consuming. Its advantage is that after compilation, the conversion of the instructions into a binary sequence for use in the processing unit, the execution run-time is minimized. In the routine described below the PC waits to retrieve a block of 128 data values which could, for example, be the frequency spectrum produced from a fast Fourier transform. These data values are generated by a separate microprocessor mounted on the specialist processing board. (The board would also have it own algorithms and memory.)

```
. . . . . . . . .
pascal instruction
pascal instruction
pascal instruction

inline(
        $BA/$0294/              {          MOV   DX,294H}
        $B3/$20/                {          MOV   BL,32}
        $EC/                    {LOOP 1:   IN    AL,DX}
        $84/$C3                 {          TEST  AL,BL}
        $74/$FB                 {          JZ    LOOP1}
    );
N: =1;
REPEAT
        inline (
                $BA/$0294/      {          MOV   DX,294H}
                $B3/$02/        {          MOV   BL,2}
                $EC/            {LOOP 2:   IN    AL,DX}
                $84/$C3         {          TEST AL,BL}
                $74/$FB         {          JZ LOOP2}
            );
DATA[N]: = PORTW [$290];
CLEAR: = (PORT[$294] AND 2);
N: = N+1;
UNTIL N=129;

pascal instruction
pascal instruction
pascal instruction. . . . . . . . . .
```

In the first loop the PC input port 294H is checked by the PC to see if the board is ready to transmit data to it. The board will set a *flag* (a specific binary value) at this port when it has completed the algorithm that generates the 128 values. When this is the case the PC moves on to the second test (loop 2). Then each time the board moves a data value to its output port it again sets the flag. This triggers the PC to retrieve this data value presented to it at its input port address 290. After collection the PC then resets the location 294H and loop 2 is repeated awaiting the next data value. After all 128 values have been transferred the PC returns to loop 1 waiting for the completion of the next sampled window being processed by the board. Data transfer is done in parallel format for maximum transfer speed.

Selection criteria

When selecting a data acquisition or processing board it is useful to consider the following parameters.

- The sampling rate required. This must be at least twice the frequency of the highest frequency expected in the input signal.
- The voltage resolution required and hence the number of bits.
- The number of channels needed.
- The range of the input analogue voltage expected.
- Whether the input analogue voltage range of the board can be altered.
- Whether the input analogue voltage expected is single or bipolar.
- Whether differential input may be required/is available.
- Whether a choice between a.c. and d.c. input coupling is available.
- Whether digital input/output is needed.
- Whether analogue output is needed.
- Noise levels.
- Linearity.
- Temperature dependency.
- Cost.

5.5   Smart sensors

Many sensors are now sold which have a microprocessor contained within the sensor unit and are described as *smart* sensors. This has enabled a higher specification of the sensor units to be achieved because of the on-board processing. A general scheme following the suggestion of Brignell and Dorey (1983) is shown in figure 5.14.

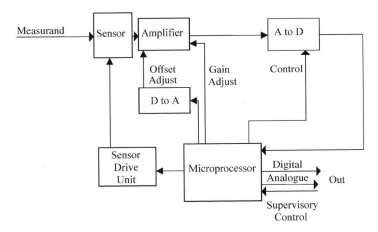

Figure 5.14. A smart sensor architecture.

The sensor itself is under microprocessor control, enabling the excitation levels to be adjusted so that switching between ranges can be done automatically. Similarly, ambient temperature and other parameters can be monitored and any corrections to the sensor output made via the analogue amplifier or at the output using data stored in ROM. The most valuable aspect of the microprocessor is the ability to produce a linear response from a nonlinear sensor using a *look-up* table stored in ROM. This has enabled the development of sensors with extremely low nonlinearity.

The output may be both digital and analogue. Analogue systems for process plant applications feature *live zeros* so that a broken line is not read by the controller as a value of zero. The standard working range is 1–5 V or, more commonly for better interference immunity on long runs, 4–20 mA. Digital telemetry is becoming more established but, at the time of writing, no standard protocol has yet been finally agreed for this Fieldbus. The supervisory control input enables the device to be part of a hierarchical control structure as well as for local operation. In addition to the provision of alarm signals to indicate when set points have been exceeded, on-board RAM can be used for data logging which can be poled later by the main controller. When used on hazardous plant, where there may be risk of explosion, a Zener diode safety-barrier device is often inserted into the electrical link. These are commercially available devices. They are mounted at some convenient place, usually at a common location for all such devices, and are part of the plant bond; see Measurement Technology (1989). Their function is to limit the flow of energy into the hazardous area, even under fault conditions, to that below what is required for ignition. Care must also be taken to ensure that plant instrumentation systems do not store sufficient energy to cause ignition. Plant safety is part of criminal law and specialist advice is essential.

Many manufacturers now market transducer units with smart features. An example is the Smart Deltapi KS Series of differential pressure transmitters by ABB Kent-Taylor. These are for use in the process industries. The addition of microprocessing power to sensors will continue to accelerate in the coming decade because of the competitive edge on sensor performance that can be achieved. There have been government funded initiatives to support SMEs (small-to-medium-sized enterprises having <250 employees) wishing to go down this route with their more traditional products.

Microprocessor developments

In order to increase processing speed, the current trend in microprocessor development has been to reduce the number of available program instructions that sequence the operations in the processor and to have the hardware execute this reduced set very quickly and efficiently. Such microprocessors are described as having reduced-instruction-set or *RISC* architectures. A

RISC device will have typically fewer than forty instructions, whereas a more traditional microprocessor might have several hundred. A further trend has been to include more memory, ADC and DAC within the microprocessor chip so that a small, stand-alone system can be easily fabricated requiring minimal additional components. Such chips are referred to as microcontrollers. Manufacturers provide development systems for their microcontrollers to facilitate program design and to emulate the system performance. These development systems run on a PC. Once the emulation is operating satisfactorily the program instructions can be downloaded and *burned* into the memory of a microcontroller chip. This is achieved using a small development board unit that connects to the PC output port. The microcontroller is then inserted into its own printed circuit board creating an independent, task-specific, stand-alone system.

A number of manufacturers now market microcontrollers. Well-known in the field are Motorola Inc, who market several series, and Microchip Technology Inc with their PIC family. Dedicated systems of this type have a wide range of applications. For instrumentation use, as well as being very suitable for inclusion in smart sensors, they can also be used to control PC plug-in boards, data loggers and dedicated instrumentation systems.

Despite the versatility and cost effectiveness of microcontrollers, development work still continues in specialist microprocessors for digital signal processing (dsp) applications. Texas Instruments Inc (TI), for example, have a long pedigree with their TMS320 series. The instruction sets of dsp microprocessors contain specific functions which are the backbone of dsp algorithms. With dsp microprocessors high performance digital filtering, large-point, fast Fourier transforms (FFTs) and correlations can be computed in real-time.

To conclude this section is a brief overview of the current trends in chip design philosophy and technology. Purely digital logic chips have been available for a decade that the user can customize directly. These are termed programmable logic arrays (PLAs) and are used to create a relatively simple device. The cheaply-produced blank chips contain of order 100 logic elements or cells. The development environment operates on a small workstation. The user creates some bespoke logic structure which is tested by simulation and then *burned down* to hard-wire the blank. In this way a small company can manufacture digital chips in-house. An example of an application is that of a burglar alarm system chip. User-defined analogue chips are, at the time of writing, beginning to be marketed. In a similar way these have a number of analogue cells that can be interconnected by burning down a design developed from the user's workstation.

More complex structures using a mix of analogue and digital signal technology can be built using ASICs (application-specific integrated circuits). These are designed by the user using software supplied by the chip manufacturer. However, manufacture has to be carried out at the silicon foundry. It is a much more expensive procedure and is only cost-effective when large volumes of chips are to be produced. Generally, when producing a traditional printed circuit board carrying analogue and digital signals it is good practice to isolate the two areas using separate ground planes and to route the connections so that high-frequency clock signals and other digital pulses do not couple through into the analogue lines. This cross-coupling can be a serious problem when designing mixed signal chips. Also, the libraries provided by ASIC manufacturers will not contain highly specialist microprocessors or operational amplifiers. Radio frequency analogue components are also unlikely to be available.

To summarize, currently the design of high-performance instrumentation systems meaning high-speed, high-resolution, performing complex signal processing with low noise and interference requires the use of discrete, individually-specified chips.

5.6   Artificial intelligence in instrumentation

There are many situations that require the processing and analysis of the information from several sensors. This may be because the operator cannot assess all the data provided at the desired speed. An example of this could be that of the pilot of a fighter aircraft where some form of *data fusion* is performed by an expert system. A different problem arises when the desired analysis may involve an assessment of the data provided by a number of sensors whose individual values may be acceptable within pre-scribed interrelated tolerances. A good example of this is the assessment of the quality of a food product where a number of parameters such as colour, pH, smell and sugar content have to be within limits but in a dependent way. For example, a higher pH may be acceptable if the sugar content is higher. It is this latter type of application that has stimulated the development of systems using artificial intelligence based on fuzzy logic and on artificial neural networks.

The definition of the expression artificial intelligence, first coined by John McCarthy in the 1950s, is like human intelligence somewhat vague. Also, as in the study of human intelligence, with the growth in understanding and the establishment of new developments, subsets of the field have been defined. The following sections briefly describe some aspects of artificial intelligence appropriate to instrumentation applications.

5.6.1   Rule-based expert systems

An expert can be defined as a person who has extensive knowledge in some field or domain. This knowledge is in two forms, *facts* and *rules*. The rules provide a structure as to how the conclusion is to be deduced from the facts. This structure is incorporated in an *inference engine*. The inference engine in its simplest format is a set of IF–THEN decision statements. The structure and sequence of these statements denotes how the expert system reaches a conclusion. The actual path taken through the structure in a given trial is referred to as the *inference chain*. There are two types of inference chain, *forward* and *backward chaining*. In backward chaining the inference structure works backward from a given hypothesis and tries to verify its validity by gathering appropriate information. An example of this procedure could be the confirmation of the suspicions of a detective following the gathering of appropriate evidence. Forward chaining begins with the information presented and attempts to draw a conclusion. Determining the cause of death at a post-mortem from signs on the body and the results of tests performed on specimens is an example of forward chaining. It is possible to have a mixed system employing forward and backward chaining which is the way much human reasoning is done. But for instrumentation applications forward chaining is likely to be the more appropriate. An additional feature required with forward-chaining is a strategy to deal with *conflict resolution*. A conflict occurs when more than one rule *fires* as the inference engine works on the data. The resolution is achieved by using a strategy such as prioritizing the rules, or not using a rule after it has scored a hit, or using the rule that relates to the most recent addition to the data memory.

The earliest commercial applications of AI came from the need to incorporate the expertise of a renowned specialist or the combined skills of a team into a knowledge base that could then be used either by a novice or be available remotely or be processed at a much higher speed than is humanly possible. It is easy to imagine examples of applications of such expert systems.

- A medical expert in the diagnosis and treatment options of some form of cancer could have their lifetime of experience made available on-line, world-wide.
- The safe operation of a remote engineering installation such as a drilling platform or space vehicle. The skills of the designers and technicians could be stored on-board so that when certain sensors produced alarm warnings the diagnosis and remedy of the situation could be obtained even if communications to base were interrupted.
- The fault-diagnosis of some complex device following its production or for servicing. A good example of this is the growing use of computer diagnostics for vehicle servicing. Long-range transportation companies could log the state of their vehicles on the road for analysis on return to the depot. In this way it should be possible to anticipate expensive breakdowns a long way from home.

- The complexity of modern aircraft means that the flight crew may not be able to react fast enough to the warnings of an attack or not have enough skill to appreciate the significance and appropriate remedy from a set of alarm signals. The use of expert systems on commercial airliners has enabled a reduction in the number of flight crew required.

Durkin (1994) contains a survey on the application fields of expert systems.

It is generally accepted that rule-based expert systems are the best approach to producing a knowledge-based system, Negnevitsky (2002). Expert systems are tailored to the application. But the general design, as outlined above, is to have a large knowledge-base from which specific conclusions can be inferred. This may involve prompting the user for more data but will ultimately involve matching the input data with specific states or conditions within the knowledge-base. A probability of match might also be issued. In some applications an automatic response may be output. For example, when the forward and ground radar of a high-speed, low-flying, fighter aircraft detect a hill ahead. This is a control application rather than instrumentation. Much work has been done and research still continues apace on the applications of AI in control theory, particularly in the field of robotics.

Programming languages have been specifically designed for expert system use. The most well-known are LISP, PROLOG and OPS. LISP was defined by John McCarthy and is the oldest. The cheaply available computing power that became available in the 1980s with the PC and other workstations has led a number of software houses to market packages for expert system development. Consequently, the user does not need to become a proficient programmer in one of these languages. The development environment is referred to as the *shell*. This is a software package containing all the relevant utilities so that the user has only to input the facts and rules and to design at the structural level.

5.6.2 Fuzzy logic

The description of a system, problem-solving and decision-taking is seldom achieved using simple concepts of black–white but more of setting value judgments to various shades of grey. For example, when describing the temperature of a water tank expressions such as cold, cool, warm, hot, very hot are used. This is in contrast to the simple machine approach of defining a threshold temperature and describing the state of the water as hot or cold depending on which side of the line it fell.

In an attempt to incorporate linguistic descriptors into a mathematical representation the Polish logician and philosopher Jan Lukasiewicz in 1930 proposed the concept of multi-valued logic. The logic values could take any real number between 0 and 1. The actual number assigned to a statement was the possibility of it being true or false. This became known as *possibility theory*. Max Black (1937) suggested that the actual numbers assigned to states in a discrete, multi-valued logic system could be derived from the opinions of a group of people. He was the first to define a simple *fuzzy* set. In a paper Lotfi Zadeh (1962), Professor in Electrical Engineering at Berkeley, stated that a new kind of mathematics was needed for these fuzzy quantities. Further to this he argued that traditional probability distributions were not appropriate when attempting to incorporate natural language into system theory. Three years later Zadeh (1965) published a paper defining a formal system of mathematical logic for fuzzy set theory. Initially he believed that fuzzy logic would find its home in the humanities but in this latter paper he suggested that it would play an important role in control engineering.

A *set* is a classification grouping. In natural language classification is made according to whether something belongs to a particular set. A cow belongs to the animal set. A Rolls-Royce belongs to the car set. In traditional set theory membership would be in or out, TRUE or FALSE. Membership of the set would be assigned binary logic 1. Non-membership would be assigned logic 0. Sets defined in this stark way are sometimes described as *crisp*. The dynamic range of the variable in question, such as the temperature of a tank, is described as the *universe of discourse*. For a particular universe of discourse there could be a family of sets. In a crisp set having a universe of discourse X where a particular value of x may or may not belong, the notation used is

$$\text{if } x \text{ belongs to } X \text{ then,} \quad x \in X$$

$$\text{if } x \text{ does not belong to } X \text{ then,} \quad x \notin X.$$

Zadeh's concept was to have graded membership. A fuzzy set called A has a *membership function* $\mu_A(x)$ which characterizes the *degree of membership*. The defining statement for the membership function is

$$\mu_A(x) : X \rightarrow [0,1]. \tag{5.37}$$

The membership value of a particular point x is in the range

$$0 \leq \mu_A(x) \leq 1. \tag{5.38}$$

Figure 5.15 relates to the temperature of the tank described earlier. A crisp set and a fuzzy set are shown for the descriptor *hot*.

For the tank example given above there is a family of fuzzy sets: cold, cool, warm, hot and very hot. It is possible to have a fuzzy set contained within another set on the same universe of discourse. These are sometimes referred to as *fuzzy subsets*. For example very hot might contain the subset very, very hot. A crisp subset must be entirely contained within the set to which it belongs. However, fuzzy sets can spill over. When dealing with subsets instead of describing the dynamic range as the universe of discourse

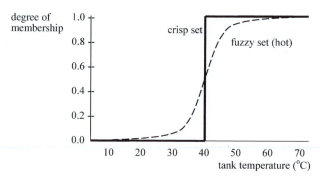

Figure 5.15. Crisp and fuzzy sets.

the notion of a *reference superset* is sometimes used. All values x_i in the reference superset must have degree of membership 1. In other words, a reference superset is a crisp set where $\mu_X(x_i) = 1$.

To create a fuzzy set it is necessary to define the membership function μ. If the universe of discourse is discrete then for each point in the range a value is needed for its degree of membership to that set. An expert system can be created by asking the expert to estimate the degree of membership of values belonging to some universe of discourse to a fuzzy set. These numbers can then be entered into a computer program as a *vector*. For example, with the tank of figure 5.15 if the discrete temperature values are

$$X = 10, 20, 30, 40, 50, 60, 70 \quad [^{\circ}\text{C}]$$

then the fuzzy set hot might be estimated as having a membership

$$\text{hot} = (0.15 \quad 0.17 \quad 0.19 \quad 0.5 \quad 0.9 \quad 0.93 \quad 0.95).$$

Commonly, fuzzy set vectors are written so that a particular membership function value has with it that universe of discourse value to which it belongs. The symbol '/' is used to simply separate the membership value from the variable value. So, in the example given,

$$\text{hot} = (0.15/10 \quad 0.17/20 \quad 0.19/30 \quad 0.5/40 \quad 0.9/50 \quad 0.93/60 \quad 0.95/70).$$

Membership functions can also be input into a fuzzy logic program as a continuous mathematical function. Commonly used are the sigmoid, Gaussian and pi functions. For a mid-range fuzzy set a triangle shape is often used (figure 5.16). Commercial software for developing fuzzy logic systems usually contains these functions as standard options. The user simply enters some parameters to define the exact shape of the function type.

It is possible to further shape a given fuzzy set to incorporate natural language. So, descriptors such as *very* and *extremely* reduce the weighting of the mid-range of the fuzzy set, whereas expressions such as *moderately* and *fairly* enhance the mid-range. These fuzzy set modifiers are known as *hedges*. Figure 5.17 illustrates this concept on the fuzzy set cold.

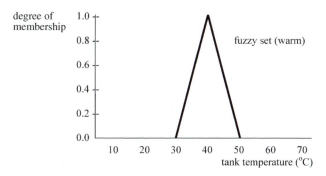

Figure 5.16. A triangular fuzzy set.

The dashed line is obtained by squaring the function that generated the sloping line. The dashed line would be described as *very* cold. The dotted line is obtained by square-rooting the function that generated the sloping line. This might be described as *moderately* cold. In general, changing the membership function to $(\mu)^n$ where n is a positive value will concentrate $(n > 1)$ or dilate $(n < 1)$ the membership values of the fuzzy set by whatever degree is required. Asymmetrical functions can also be easily created and used such as a mid-range triangle but with one side hedged into a curve.

To generate the output of a fuzzy logic system logic operations are performed on the data. Georg Cantor developed classical set theory towards the end of the 19th century. In the fuzzy logic sense this is crisp set theory. Three classical set operations are complement, intersection and union. Their performance on fuzzy sets is described below. Whenever an operation is performed a new set is created.

Complement

This is the logic inverse or NOT operation, so performing the complement operation on the fuzzy set hot would generate a fuzzy set NOT hot. The

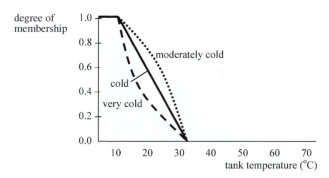

Figure 5.17. Hedging of a fuzzy set.

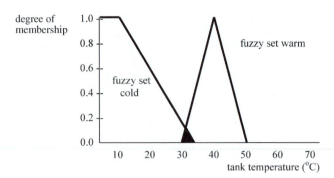

Figure 5.18. The intersection operation.

mathematical statement is

$$\mu_{\text{NOT hot}} = 1 - \mu_{\text{hot}}. \tag{5.39}$$

Intersection

In crisp set logic this is simply an AND operation. With fuzzy sets the smallest values within the overlap have to be extracted. It is sometimes called a *min(imum)* operation. In the example of figure 5.18 the dark triangle is the intersection, which might be called *tepid*. The notation used to represent this operation is ∩. So in the example given the new set is

$$\mu_{\text{tepid}} = \mu_{\text{cold}} \cap \mu_{\text{warm}}. \tag{5.40}$$

Union

In this operation values are taken from one or both fuzzy sets. It is the logic INCLUSIVE OR operation. The largest value is used and so it is sometimes referred to as a *max(imum)* operation. Again, a new set is created. In the example of figure 5.19 it might be called *not tepid*. The symbol used is ∪. So

$$\mu_{\text{not tepid}} = \mu_{\text{cold}} \cup \mu_{\text{warm}}. \tag{5.41}$$

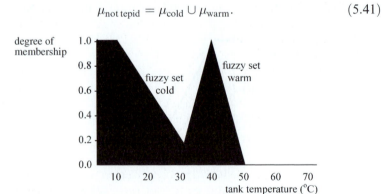

Figure 5.19. The union operation.

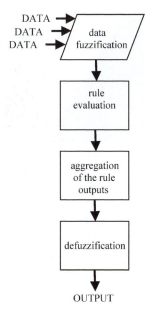

Figure 5.20. The Mamdani inference method.

Finally, having described fuzzy sets and logic operations, the overall procedure is examined. The task of the fuzzy system is to combine the data input and produce a crisp output value. In fuzzy set theory the process of interpreting various inputs into output(s) is termed *inference*. Mamdani and Assilian (1975) published a paper on the use of a fuzzy system for the control of a boiler and steam engine. The inference technique employed is now the most frequently used and is referred to as the Mamdani method. Figure 5.20 shows the strategic stages.

(1) Fuzzification

The first stage is to take the specific input data values and to determine their membership value within each appropriate fuzzy set. As previously explained, an expert on the tank system has already estimated the degree of membership values for the sets. For example, referring to figure 5.21, the input temperature data value of 27 °C has a membership of 0.3 in the fuzzy set cold, 0.0 in the fuzzy set warm and 0.05 in the fuzzy set hot. (In fuzzy logic terminology numerical input data are said to be crisp.)

(2) Rule evaluation

The *rules* are the equivalent of IF–THEN statements in binary logic. The logic operations described above are how the rules are implemented. Fuzzy rules were developed in a further paper by Zadeh (1973). To illustrate the

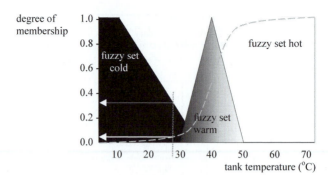

Figure 5.21. The tank temperature fuzzy sets.

process, in the water tank example the pressure in the tank could be included in the problem as a second data input. A new universe of discourse is now introduced which could have its own fuzzy sets of *low pressure, normal pressure, high pressure*. This is illustrated in figure 5.22. The water tank expert who has a great deal of practical experience of the interrelation of the temperatures and pressures and their control states that

Rule 1

If the temperature is **COLD and** the pressure is **NORMAL** then the heater is ON.

Rule 2

If the temperature is **HOT or** the pressure is **HIGH** then the cooler is ON.

These two rules (more of course are possible) each have a corresponding set operation **and**, **or** and each produces a new fuzzy set. In the example given the inputs are that the temperature is 27 °C and the pressure is 7 atmos (gauge).

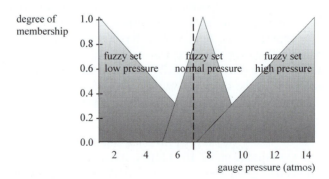

Figure 5.22. The tank pressure fuzzy sets.

(3) Aggregation of the rule outputs

In figure 5.23 the grey triangle is the result of rule 1. Here the apex of the triangle (degree of membership 0.30) is determined by the fuzzy set cold for the temperature 27 °C. This is because the pressure of 7 atm in the fuzzy set normal pressure has a degree of membership of about 0.75. The **and** operation returns 0.3 as the common minimum value.

The black cusp shape below the curve is limited by the hot fuzzy set. Here 27 °C has a membership of 0.05. Rule 2 is an **or** operation but the pressure of 7 atmos has a membership of zero in the fuzzy set high pressure and so has no effect on this result. The value 0.05 is the maximum degree of membership from either of the two sets.

(4) Defuzzification

Once the logic operations have been performed and the results aggregated the system needs to deliver a single, crisp output. A useful feature of figure 5.23 is that it gives a visual impression of the weightings of each rule output in the aggregation. The most common way that a crisp output is generated is from the centroid, shown as $\otimes$ in the figure. The centroid is the centre of mass for some three-dimensional object. The centre of mass is that point around which the sum of all moments is zero. To find the x coordinate for this point x_{cm} in an object having discrete elements of mass m_i equation (5.42) is used:

$$x_{cm} = \frac{\sum_i m_i x_i}{M}.$$
(5.42)

Translating this into values of figure 5.23 gives

$$\bar{v} = \frac{\sum_{i=a}^{b} \mu_i v_i}{\sum_{i=a}^{b} \mu_i}$$
(5.43)

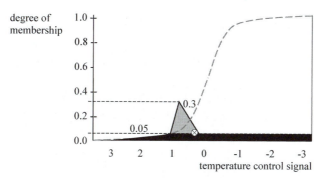

Figure 5.23. Aggregation of the two rule outputs for 27 °C at a pressure of 7 atm.

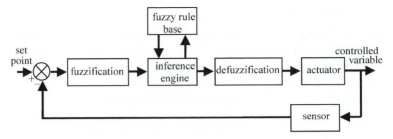

Figure 5.24. A simple fuzzy logic-controlled process.

where $\bar{v}$ is the crisp output control signal and i is a particular value in the universe of discourse v which extends over the range a to b.

Zadeh's prediction about an important application of fuzzy logic being in control engineering has in fact come true. Figure 5.24 shows a block diagram for a simple control system.

Because the skills and knowledge of an expert operator can be incorporated into the controller it has been found that nonlinearities, disturbances, drift and sensor degradation can be successfully accounted for. Because of the nature of fuzzy logic the skills and knowledge of the expert can be input in linguistic terms. A formal mathematical model is not needed. Another advantage of fuzzy logic control is that modification to the skills or rule-base can be made very easily while in use.

McNeill and Freiberger (1993) give a history of early industrial applications of fuzzy logic. The first commercial application of fuzzy control was for a Smidth-type cement kiln; see Holmblad and Østergaard (1982). Cement is a mixture containing the complexes calcium aluminate, calcium silicates and calcium alumino-ferrite. Clay, limestone, sand and iron ore are input as slurry. The kiln is a rotating drum approximately 5 m in diameter and 140 m long revolving at about 1 rpm. It operates at a temperature around 1300 °C and the through-time is in the region of 3 h. The variables to be controlled are the fuel supply, slurry feed and drum rotation rate. Temperature control is critical to forming the correct quality of granules, called *clinker*. Previous attempts at computer control had failed. It was the discovery of 27 IF–THEN rules contained in a training manual for the kiln operators that led the researchers to develop a fuzzy verbal rule-base. A typical rule might be

'if the oxygen percentage is rather high and the free-lime and kiln drive-torque are normal decrease the flow of gas and slightly reduce the fuel rate'.

A successful trial was carried out in 1978. This was the first industrial use of a closed-loop, expert-system controller. A significant proportion of cement kilns world-wide now use fuzzy logic controllers.

In principle any number of inputs and outputs can be incorporated into a controller. However, with systems having many input–output variables the rule structure can become very complex. For such systems it is often easier to develop a number of fuzzy logic controllers to handle sub-systems in the process.

Other areas where fuzzy logic has been applied are

- information retrieval,
- navigation systems,
- timetabling for transportation systems,
- a number of fields in robotics,
- feature definition in imaging systems,
- financial planning,
- meteorological analysis,
- predictive control for the automatic operation of a train,
- domestic electrical goods.

5.6.3   Artificial neural networks

In section 5.6.1 the rule-based expert system was described. This is based on formal logic using IF–THEN statements. The fuzzy logic of section 5.6.2 can be seen as an extension of this using logic which is less 'binary' by nature. However, artificial neural networks do not use logic. They require no prior knowledge or skills from an expert to be programmed into the sotware. They are systems that attempt to mimic the learning processes of the brain and, as such, learn the relationships between data input and output. It could be argued therefore that artificial neural networks show true intelligence. This is in contrast to the machine-intelligence of the expert systems where intelligence is installed by the designers.

The fundamental building block is a mathematical model of the biological neurone. These artificial neurones are grouped into some structure that is then trained using sets of data. The software *learns* the relationships needed to fit the data and can then be used on unseen data to derive a conclusion. Once trained a hard-wired version may be built if desired. The following sections describe the essential features of how the artificial neural network is constructed, trained and used.

The biological neurone

The neurone is the fundamental cellular unit of the nervous system and the building block of the brain (figure 5.25). Neurones occur in a number of forms depending on their function and location in the body but have a similar overall structure.

Each neurone is a processing unit that receives signals from many other neurones through its input structure of *dendrites*. If the combined input

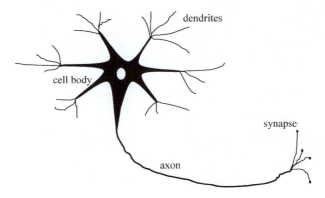

Figure 5.25. A biological neurone.

stimuli reach the *threshold level* the cell is triggered and a signal passes on to the *axon* and through to the output structure leading on to many more neurones. Its nearest logic equivalent is the AND gate but the system is not simple binary.

The axon has a high-resistance outer membrane and contains a low-resistance fluid called *axoplasm*. It is the modification of cation concentrations across the membrane that results in a voltage wave passing along the neurone.

Table 5.3 shows an imbalance or ion differential across the membrane that is maintained against diffusion and Coulomb forces by *active transport*. Carrier molecules restore any disturbance to this imbalance using energy supplied from the respiratory system. The net difference in ion concentrations between inside and out, of 144.4–116 mmol per litre, produces a potential difference of between 60 and 90 mV. The membrane is said to be *polarized* with this *resting potential*. If the neurone receives a sufficiently high stimulus from other input neurones or external stimuli such as heat, physical stress, electric current or chemical intervention, the cell wall can become permeable to the ions. An external stimulus to the neurones in the brain is not the normal method of creating activity! Instead, the input is derived from neurones in the rest of the body or from other neurones in the massive neural tree-structure within the brain. Once the potential difference drops below the *action potential* in some region (the actual drop

Table 5.3. Ion concentrations in and around the neurone.

Concentration of ions	Sodium ions (Na^+) (mmol per litre)	Potassium ions (K^+) (mmol per litre)
Outside cells (tissue)	140.0	4.4
Inside cells	16.0	100.0

in potential difference may be as low as zero or even go negative) this triggers adjacent parts of the membrane to become permeable and so cause a voltage pulse-wave to travel down the axon. After the impulse has passed a recovery, or *refractory*, period follows of about 1 ms during which the ion molecular pumps return the cation concentration levels to the resting state of table 5.3.

It was Helmholtz in 1850 who first measured transmission speeds along nerve fibres. Initially it was thought that these electrical pulses travelled at the same speed as electrical signals along wires. But instead of the expected $2 \times 10^8 \, \mathrm{m\,s^{-1}}$ he measured speeds of only $25 \, \mathrm{m\,s^{-1}}$. Some neurones in the body, particularly those involving the reactions, have an insulated sheath around the axon called *myelin*. The sheath has gaps at various positions along its length to aid re-polarization. In this type of neurone transmission speeds may be up to $100 \, \mathrm{m\,s^{-1}}$.

The transmission between neurones is through the *synapses*. Biochemical agents called neuro-transmitters are released by the axon of the transmitting neurone and accepted by the dendrites of the receiving neurones. If the incoming signal to a receiving neurone is strong enough this neurone will trigger and subsequently send pulses to further neurones. Neurones can have between 10^3 and 10^4 synapses.

The human brain contains of order 10^{11} neurones. Recent investigations have revealed that the doubling in size of the infant brain is due to growing interconnectivity of neurones rather than the growth of more neurones. Generally throughout life whenever some new information is learned by the brain new links between neurones are formed. As part of the reinforcement the synaptic efficiency is modified during the learning process. If the learning process is repeated several times with a suitable intervening period the modification to the synapse is long-term. This phenomenon is responsible for the differentiation between short-term and long-term memory.

History of artificial neural networks (ANN)

The beginning of the development of artificial neural networks can be traced back to work done in the 1940s; see Neuralware (1989). At this time researchers were attempting to understand the operation of the learning process in the biological neurone. Donald Hebb suggested that when the synaptic input and output of a neurone were both active the strength of the incoming connection was reinforced. McCulloch and Pitts (1943) created a mathematical model of a brain cell where the probability of triggering depended on the level of the input signal and the threshold at the synapse. In 1951 Minsky and Edmonds, inspired by the ideas of McCulloch and Pitts, constructed a *learning machine* out of thermionic valves, motors, clutches and a gyropilot. This device was able to learn sufficiently well to follow a maze. Following this Frank Rosenblatt developed the concept of the *perceptron*. This was an attempt to model the pattern recognition abilities of the eye. The outputs from a matrix of sensors were hard wired into an array so as to recognize certain features. In the early 1960s Widrow devised the *adaline* (adaptive linear neuron), a threshold logic device and he and Hoff developed a learning rule for correcting the connecting strengths or weightings between the input values and the summing unit. A multi-layer adaline, the *madaline*, was also developed. Minsky and Papert (1969) published a book on the perceptron and concluded that it would not be capable of learning by itself. Interest in artificial neural networks then waned for some time.

In 1978, through an association with MIT, John Hopfield of Bell Laboratories became interested in the computational abilities of biological systems. Hopfield devised a network from a study of the neuro-physiology of the garden slug. As a physicist he realized that stability in the feedback structure could be understood in terms of a minimum energy state. The learning principle in the Hopfield network is that connections between the artificial neurones are reinforced if the output of a processing element is the same as its input and, conversely, weakened if it differs. An application of Hopfield networks that has been extensively investigated is in optical pattern recognition systems.

The credit assignment problem

As mentioned in the introduction, an artificial neural network system, unlike an expert system, is able to learn the required expertise itself by training on known data. Having been trained, it can then be used on unseen data to generate novel output.

In the training process the strengthening or weakening of the inter-connections between neurones uses a procedure involving *credit assignment*. Devising a method for adjusting the synaptic efficiencies of all the neurones in an artificial neural network during its learning stage was a major challenge. Some among the early workers thought it an impossible task. The so-called *credit assignment problem* is the problem of knowing how to determine the significance a particular interconnecting weight has on the system *global error*. Credit assignment is not a problem when dealing with a single array of neurones since input and output values and appropriate weightings are easily handled. The problem arises when there are hundreds or thousands of neurones all interconnected in a large network. Only those few inter-connections at the input and output interfaces can be readily assessed for their contribution to the global error. The major challenge is how to change the weighting of a particular interconnection hidden deep within the network having only the global error to work from.

Of the various techniques developed for solving the credit assignment problem the back-propagation method has become the most popular. The development has been attributed to Rumelhart and McClelland (1986) and also to Parker (1985). The popularity of the technique is due to its success in functioning well with nonlinearly separable systems. Its success did a lot to revive interest in artificial neural networks and, with the rapid developments of cheap computing power, this is still continuing. The back-propagation process is described below. Elaborations are described in Chauvin and Rumelhart (1995).

Mathematical structure for the back-propagation technique

The multiple sets of equations needed to describe the structure requires some rather cumbersome notation that is set out below. The artificial neural network structure shown in figure 5.26 gives the labelling for the artificial neurones ⊗ and their inter-connecting weights w.

All the artificial neurones are inter-linked from the input upwards although for clarity these are not all shown. A route through the hidden layers is shown as ith to jth to kth neurones with their interconnecting weights w_{ij} and w_{jk}. Different numbers of neurones are shown in different layers to indicate that no set pattern is required. Generally, it has been found that three hidden layers are the useful upper limit. Adding more layers does not tend to improve performance and increases the learning time.

The input to the jth neurone in level s is $I_j^{[s]}$ and is the sum of the variously weighted outputs of all the neurones in the previous layer,

$$I_j^{[s]} = \sum_i w_{ij}^{[s]} X_i^{[s-1]}.$$ (5.44)

The output of the jth neurone in level s is $X_j^{[s]}$ and is some function f of its input $I_j^{[s]}$,

$$X_j^{[s]} = f(I_j^{[s]}).$$ (5.45)

Various functions have been tried for f such as sine and tanh but the one which has become the most popular closely resembles the biological neurone and is called the sigmoid function,

$$f(I_j^{[s]}) = \frac{1}{1 + \exp(-I_j^{[s]})}.$$ (5.46)

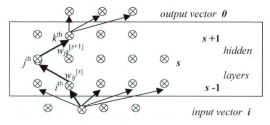

Figure 5.26. A simple artificial neural network structure.

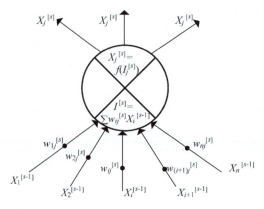

Figure 5.27. An artificial neurone.

A very useful feature of this function is the efficient way its derivative can be calculated,

$$\frac{\partial f}{\partial I_j^{[s]}} = f(1-f). \tag{5.47}$$

The input–output terms for the artificial neurone are shown in figure 5.27. Initially, all the weights are set at random. For the learning process to proceed a relationship is needed that relates the value of a particular weight w to the global error, E. Then the weight can be adjusted by some amount Δw. To achieve this the so-called *gradient descent rule* was defined,

$$\Delta w_{ij}^{[s]} = -\text{lcoef} \frac{\partial E}{\partial w_{ij}^{[s]}} \tag{5.48}$$

where lcoef is a learning coefficient. Equation (5.48) is the essence of the learning process in that it is the technique by which the global error is apportioned to all of the connecting weights. It is a way of addressing the credit assignment problem discussed above. The partial differential term is a measure of how sensitive the global error is to a particular weight. The more influential a weight is the more it will be modified during the training process. The learning coefficient is a value chosen, somewhat arbitrarily, between 0–1. If it is too large the system may go into *oscillation* during training. The changes to the weights are too large and the system cannot converge to the minimum value of E. If it is too small the learning times will be longer and there is a greater probability that the system will settle at some minima that is not the lowest possible value of E.

From equation (5.48) it can be seen that it is a requirement that E should be a function that is differentiable with respect to each of the interconnecting weights throughout the network. A squared function is used,

$$E = \frac{1}{2} \sum_{\substack{\text{output} \\ \text{neurones}}} (X_d - X_a^{[o]})^2 \tag{5.49}$$

following the concept of a *least squares* analysis where X_d and $X_a^{[o]}$ are the presented desired output and the calculated actual output respectively for each neurone in the output layer. (The superscript [o] signifies the output layer.) The factor $\frac{1}{2}$ is simply for convenience of normalization and becomes 1 after differentiation. The successful development of the back-propagation technique was in the concept of first calculating the error in the output layer and then assigning this appropriately to all the hidden neurones throughout the network. The proportion of the error ascribed to an individual neurone is determined using the *local error* of the neurone calculated from the partial derivative of E. The local error δ of the jth neurone in level s is determined using

$$\delta_j^{[s]} = -\frac{\partial E}{\partial I_j^{[s]}}. \tag{5.50}$$

The use of the partial derivative means that δ is the sensitivity of the global error E to the input I of that particular neurone. The usefulness of this definition for the local error can be seen by expanding the second term in equation (5.48) using the chain rule

$$\frac{\partial E}{\partial w_{ij}^{[s]}} = \frac{\partial E}{\partial I_j^{[s]}} \frac{\partial I_j^{[s]}}{\partial w_{ij}^{[s]}} \tag{5.51}$$

$$= -\delta_j^{[s]} \frac{\partial I_j^{[s]}}{\partial w_{ij}^{[s]}}. \tag{5.52}$$

Substituting for $I_j^{[s]}$ using equation (5.44),

$$\frac{\partial E}{\partial w_{ij}^{[s]}} = -\delta_j^{[s]} \frac{\partial (\sum_i w_{ij}^{[s]} X_i^{[s-1]})}{\partial w_{ij}^{[s]}}. \tag{5.53}$$

The partial differentiation reduces this to

$$\frac{\partial E}{\partial w_{ij}^{[s]}} = -\delta_j^{[s]} X_i^{[s-1]}. \tag{5.54}$$

This means that the equation (5.48) can now be expressed as

$$\Delta w_{ij}^{[s]} = \text{lcoef}\, \delta_j^{[s]} X_i^{[s-1]}. \tag{5.55}$$

To implement equation (5.55) as an algorithm now only requires an expression for $\delta_j^{[s]}$.

For the output layer the determination of the δ values for each neurone is straightforward. Rewriting equation (5.50),

$$\delta_j^{[o]} = -\frac{\partial E}{\partial I_j^{[o]}}. \tag{5.56}$$

Using the chain rule this can be expressed as

$$\delta_j^{[o]} = -\frac{\partial E}{\partial X_j^{[o]}}\frac{\partial X_j^{[o]}}{\partial I_j^{[o]}}. \tag{5.57}$$

But the actual error $X_d - X_a^{[o]}$ for each output neurone is known. Therefore, the first term in the right-hand side of equation (5.57) can be determined as

$$\frac{\partial E}{\partial X_j^{[o]}} = \frac{\partial(\frac{1}{2}\sum(X_d - X_a^{[o]})^2)}{\partial X_j^{[o]}} \tag{5.58}$$

$$= -(X_{jd} - X_j^{[o]}) \tag{5.59}$$

where X_{jd} is the presented desired output for the jth output neurone.

Also, it can be seen from equation (5.45) that the second term in equation (5.57) is the partial derivative of f:

$$\frac{\partial f}{\partial I_j^{[o]}} = \frac{\partial X_j^{[o]}}{\partial I_j^{[o]}}. \tag{5.60}$$

By combining equations (5.59) and (5.60), $\delta_j^{[o]}$ in equation (5.57) can be expressed as

$$\delta_j^{[o]} = (X_{jd} - X_j^{[o]})\frac{\partial f}{\partial I_j^{[o]}}. \tag{5.61}$$

The partial derivative $\partial f/\partial I_j^{[o]}$ can be replaced by numerical estimates using equations (5.46) and (5.47).

The determination of the local error values for neurones in the hidden layers is more involved. A general expression is required for $\delta_j^{[s]}$ that involves the $s + 1$ layer. Once this is obtained errors occurring between the system outputs and the training values presented can be determined as described above. But then weightings can be adjusted for the interconnections between neurones in the output and in the penultimate layer using $\delta_j^{[s+1]}$. This procedure can be repeated propagating backwards using $\delta_j^{[s]}$, $\delta_j^{[s-1]}$ etc., towards the input layer of neurones. The procedure is termed back-propagation because of this working towards the input layer of neurones from the output layer.

Adapting the notation of equation (5.57) for some general level s gives

$$\delta_j^{[s]} = -\frac{\partial E}{\partial X_j^{[s]}}\frac{\partial X_j^{[s]}}{\partial I_j^{[s]}}. \tag{5.62}$$

Again, using the chain rule, the first term in equation (5.62) can be expanded to give

$$\frac{\partial E}{\partial X_j^{[s]}} = \sum_k \left(\frac{\partial E}{\partial I_k^{[s+1]}} \frac{\partial I_k^{[s+1]}}{\partial X_j^{[s]}} \right). \tag{5.63}$$

The summation over k appears because all the neurones in level $s + 1$, and consequently E also, are affected by changes in $X_j^{[s]}$. Using the definition for δ equation (5.63) can be rewritten as

$$\frac{\partial E}{\partial X_j^{[s]}} = \sum_k \left(-\delta_k^{[s+1]} \frac{\partial I_k^{[s+1]}}{\partial X_j^{[s]}} \right) \tag{5.63}$$

$$= \sum_k \left(-\delta_k^{[s+1]} \frac{\partial \sum_j w_{jk}^{[s+1]} X_j^{[s]}}{\partial X_j^{[s]}} \right). \tag{5.64}$$

Because the second term in equation (5.64) is a partial derivative in $X_j^{[s]}$, unless $i = j$ the differential is zero, so equation (5.64) becomes

$$\frac{\partial E}{\partial X_j^{[s]}} = \sum_k \delta_k^{[s+1]} w_{jk}^{[s+1]}. \tag{5.65}$$

Consequently, equation (5.62) becomes

$$\delta_j^{[s]} = \frac{\partial f}{\partial I_j^{[s]}} \sum_k \delta_k^{[s+1]} w_{jk}^{[s+1]}. \tag{5.66}$$

As before, the partial derivative $\partial f / \partial I_j^{[s]}$ can be replaced by numerical estimates using equations (5.46) and (5.47).

All the components of equation (5.65) can now be determined so that the adjustments to the weights can be made every time an input–output pair of training vectors is presented to the network. In practice, files containing training data pairs are constructed containing hundreds of values of these vectors. These are fed into the system randomly during the learning process. This random selection enables each training value to be used several times but with reduced risk of having the system settle into less than the minimum global error. A few hundred data values could typically be used to provide tens of thousands of training cycles. Depending on the network size and the computer used, training may take several hours. The training algorithm is summarized in the flow chart shown in figure 5.28.

Once training is complete the weights are left in place and the system can be used on new, unseen data. Errors in performance can be low, often down to a few percent. This is clearly a poor performance when deriving values for a system where a known, causal relationship exists or can be easily determined. However, the advantage of artificial neural networks is that the causal relationship need not be known. The network is trained to

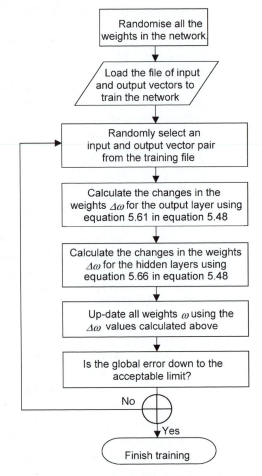

Figure 5.28. Flow chart for the back-propagation training technique of a neural network structure.

recognize input–output relationships and then provide an estimate of an output for an unseen input. The unknown relationship may even be nonlinear.

It is possible to build the trained system into a hard-wired structure using op-amps and resistors. Artificial neural network integrated circuits have also recently become available. Because such systems use analogue techniques, they can operate extremely quickly. They also have the advantage of being fault tolerant. In a remote system such as a satellite, where radiation damage is likely, the hard-wired system will degrade slowly as individual neurones malfunction. A microprocessor-based system in similar circumstances would fail catastrophically if the CPU or RAM were damaged.

Development work is continuing in training strategies and in the design of new network structures containing sub-loops and internal feedback. Some software-based systems, such as for example in control applications, have their weight values updated periodically as the plant characteristics age. Combinations of artificial intelligence techniques are also being investigated such as with fuzzy logic and expert systems.

Some fields of application are listed below.

- text to speech conversion,
- image compression,
- handwriting recognition,
- object classification for image recognition and industrial inspection,
- predictions of stock market and weather,
- system modeling,
- noise filtering,
- parameter adjustment to improve tolerances in robotic systems,
- functional synthesis.

References

Black M 1937 'Vagueness: An exercise in logical analysis' *Phil. Sci.* **4** 427–455

Brignell J E and Dorey A P 1983 'Sensors for microprocessor-based applications' *J. Phys. E* **16**(10) 952–958

Chauvin Y and Rumelhart D E eds 1995 *Backpropagation: Theory, Architectures, and Applications* (Lawrence Erlbaum Associates)

Durkin J 1994 *Expert Systems Design and Development* (Englewood Cliffs, NJ: Prentice Hall)

Holmblad L P and Østergaard J J 1982 *Control of a Cement Kiln by Fuzzy Logic, Fuzzy Information and Decision Processes* (Amsterdam, North-Holland) pp 398–399

Horowitz P and Hill W 1989 *The Art of Electronics* 2nd edition (Cambridge University Press)

Mamdani E H and Assilian S 1975 'An experiment in linguistic synthesis with a fuzzy logic controller' *Int. J. Man–Machine Studies* **7**(1) 1–13

McCulloch W S and Pitts W H 1943 'A logical calculus of the ideas immanent in neural nets' *Bull. Math. Biophys.* **5** 115–133

McNeill D and Freiberger P 1993 *Fuzzy Logic* (New York: Simon and Schuster)

Measurement Technology Ltd 1989, Application Note AN9003, *A User's Guide to Intrinsic Safety*, Luton, Bedfordshire, England

Minsky M L and Papert S S 1969 *Perceptrons* (Cambridge, MA: MIT Press)

Negnevitsky M 2002 *Artificial Intelligence: a Guide to Intelligent Systems* (Addison Wesley)

Neuralware Inc. 1989 *NeuralWorks Professional II User's Guide*

Parker D B 1985 *Learning Logic Report TR-47* (Cambridge, MA: MIT)

Rumelhart D E and McClelland J L (eds) 1986 *Parallel Distributed Processing: Explorations in the Microstructure of Cognition, 1, Foundations* (MIT Press)

Zadeh L 1962 'From circuit theory to system theory' *Proc. Radio Engineers* **50**

Zadeh L 1965 'Fuzzy sets' *Information and Control* **8**(3) 338–353
Zadeh L 1973 'Outline of a new approach to the analysis of complex systems and decision processes' *IEEE Trans. Systems, Man and Cybernetics* **SMC-3**(1) 28–44

Exercises

1. The analogue output from a certain sensor is to be converted to digital format. If a resolution of 0.5% is required how many bit ADC is needed? If the signal contains frequencies up to 1.5 kHz what should be the sampling rate? Explain your choice.

2. Choose an ADC for the following applications and explain your choice:
   (i) digital audio recording studio,
   (ii) multimeter,
   (iii) vibration sensor on an aircraft engine.

3. A manufacturer of portable audio mixing desks decides to replace the heavy, multi-core cable that runs from the desk to the stage by an optical fibre. Calculate the serial bit-rate down the fibre given that there are 32 signal channels, each with 18-bit resolution and that the maximum audio frequency is 20 kHz.

4. A certain monochrome, video-camera CCD sensor has an array of 512×512 pixels. If a brightness resolution of approximately 2% is required what bit size must the analogue-to-digital converter be? Determine the output bit-rate if the camera generates 25 frames per second. Suggest a suitable type of ADC.

5. A smart pressure gauge is to be designed based on a stainless-steel diaphragm and strain gauges; see equations (3.2) and (3.3) in section 3.2.2. Draw a block diagram of the type figure 5.14 stating what features you would incorporate in the device to achieve an overall error of <0.1%. (The phase-sensitive detector of section 6.4.2 might be a useful inclusion.)

6. An automated hot-drink vending machine is to be controlled by a rule-based expert system. A hot water tank is maintained at 70 °C by an electrical heater and conventional thermostat. The tank is connected to the water supply and a float valve maintains the water level in the tank. TEA, COFFEE, DRINKING CHOCOLATE and TOMATO SOUP are normally available. SUGAR, EXTRA SUGAR and WHITENER are options. Construct a flow chart of the logic sequences necessary to incorporate the following features:
   • If all systems are operational but no action is under way an indicator READY comes on.
   • The money-accept slot does not retain money when the system is shut down (SYSTEM FAULT).

- A water level sensor shuts off the heater and closes down the operation if the level is too low and an indicator light SYSTEM FAULT comes on to inform the customers.
- If the water temperature is low an indicator light asks the customers to PLEASE WAIT. No other functions operate.
- If the temperature rises to 100 °C the power to the immersion heater is isolated and the system shuts down. The indicator light SYSTEM FAULT comes on.
- If there are no cups left the system shuts down. The indicator light SYSTEM FAULT comes on.
- The ingredient hoppers contain sensors. If any hopper is empty an appropriate light informs the customers. If a drink is not available the selection button for that drink becomes inoperative and the money-accept slot does not retain money.
- If the coin sensor detects that sufficient coins have been entered they are stored in a holding box until the drink has been dispensed. Otherwise the coins are returned.
- If the money available in the machine to dispense change has all been used an indicator NO CHANGE AVAILABLE comes on. If excess money is input it is returned.
- If money is being stored in the holding box the selection procedure becomes operative.
- The customer can select one drink only. The selector buttons become inoperative after a selection has been made.
- Sugar and whitener buttons become operative after drink selection unless soup is chosen.
- If the cup sensor has detected that a cup is in place the dispensing begins. If not the indicator SYSTEM FAULT comes on and money is returned and then the system shuts down.
- If the system has successfully dispensed then any change due is released.
- The READY indicator comes back on.

7. An executive-car manufacturer decides to develop an electronically-controlled, automatic, 4-speed gearbox. This is expected to simplify the complexities of the gearbox hydraulics and provide greater flexibility when driven. Fuzzy logic is proposed as the best method of producing a controller. Your task is to develop some fuzzy sets for
   - each gear versus road speed,
   - accelerator pedal position,
   - level of braking applied,
   - engine rpm.
   The following rules are to be incorporated for the choice of gear to be output to the selector mechanism:
   (a) If the vehicle is stationary then select 1st gear.

(b) If the road speed is between about 5–20 mph then select 2nd gear.

(c) If the road speed is between about 15–40 mph then select 3rd gear.

(d) If the road speed is above about 35 mph then select 4th gear.

(e) If the accelerator pedal has been pressed beyond about midway then changes in (a)–(d) above should occur toward the higher road speed ranges given.

(f) If the accelerator pedal is depressed fully (kick down) then drop to the next lower gear, excluding 1st gear, and only change up when the engine speed exceeds about 4000 rpm.

(g) If the accelerator pedal is returned from kick down to midway then move to the next higher gear.

(h) If the accelerator pedal is released and the brakes are applied gently but the vehicle does not decelerate then, after several seconds, move to the next lower gear, excluding 1st gear.

(i) If the ABS system has become active change to the next higher gear.

8. (a) By reference to the biological neurone explain the operation of the artificial neurone.

   (b) Give an expression for the output of an artificial neurone in terms of its inputs.

   (c) What is meant by the expression *global error*? Give an equation.

   (d) The so-called *gradient descent* rule can be used to alter the weights. Explain this process.

9. The combustion chamber and exhaust manifold struts on a certain aircraft jet engine reach between 85 and 150 °C during flight. However, it is known that the temperature pattern on these ten struts can indicate an engine fault. It is decided to use an artificial neural network to warn of an engine malfunction.

   (a) Assuming an analogue signal of 0–5 V is output from each of the ten temperature measurement systems, suggest an artificial neural network structure that could be trained to give three possible outputs NORMAL, WARNING and MALFUNCTION.

   (b) With the aid of a flow chart describe the training process that would need to be employed in order for the neural network to function correctly.

Chapter 6

Data signal characterization and recovery

6.1 Introduction

When data signals from a sensor or from an instrumentation system are weak, either because the variable to be measured is tiny, the sensor output is poor, or the signal is attenuated in the telemetry link, recovery of the information may be difficult. This is because the signal of interest is of a similar magnitude to any interference or noise present. If the measurand is a steady value then the recovery process is relatively simple and the method adopted depends on whether it is noise or interference that is corrupting the data. However, if the measurand is time varying its recovery may be less straightforward. The most difficult situation is when the data itself is random and the signal is weak.

Both the data signal and any unwanted signal can be classified as either *random* or *deterministic*. The future value of a random signal cannot be predicted. There is no way of knowing or calculating the forthcoming time values. The only way the signal can be characterized is either by frequency analysis or by using statistical parameters. The future values of a deterministic signal can, in principle, be calculated. In practice the time-dependent formula may not be fully understood or even known, but the concept of the process being orderly and that the signal is following some pattern classifies it as deterministic. Table 6.1 gives some simple illustrations.

The presence of unwanted periodic signals is termed interference and commonly comes from power supply lines. The origins of noise are more varied and are described in section 6.4.

For a numerical description of the levels of both noise and interference the expression *signal-to-noise ratio* is used. The power ratio is used in decibel format,

$$10 \log_{10} \left(\frac{\text{data signal power}}{\text{unwanted signal power}} \right) = 10 \log_{10} \left(\frac{V_{\text{data}}^2 / R}{V_{\text{noise}}^2 / R} \right) \text{dB} \qquad (6.1)$$

Table 6.1. Examples of random and deterministic signals.

	Random	Deterministic
Data signal	Sounds from an engine bearing	Output from a tachogenerator
Unwanted signal	Noise	Interference

where R is the input resistance to the signal conditioning circuit. This is more commonly expressed in terms of voltages since signal powers are not normally measured. Assuming that R remains constant equation (6.1) becomes

$$20 \log_{10} \left(\frac{V_{\text{data}}}{V_{\text{noise}}} \right) \text{dB}. \tag{6.2}$$

6.2 Classification of data signals

Data signals can be sub-divided into two classifications: deterministic and random. Such data come from real processes which may themselves be deterministic or random. Random processes are sometimes called *stochastic*. The nature of the signal may not be identical with the process. For example the movement of molecules in a gas is random but the pressure measured and signal transmitted will be deterministic if the gas is in a steady state.

6.2.1 Deterministic data signals

As described above these signals come from sensor systems measuring processes that have some time-dependent relationship. An example is the temperature of a hot object as it is allowed to cool. This would be described as a transient phenomenon. The resonating panel on a vehicle would be an example of a complex periodic process, the signal from the vibration sensor containing harmonics and perhaps other non-related frequencies. The output of a resonator sensor, although measuring some steady variable, would be a sinusoidal signal with some harmonics of lesser amplitude.

Deterministic data signals can be subdivided as shown in the diagram (figure 6.1) and are described below, following Bendat and Piersol (1986).

Periodic sinusoidal data signal

In this signal a single frequency is present and can be easily described mathematically as

$$v(t) = v_0 \sin(\omega t + \phi). \tag{6.3}$$

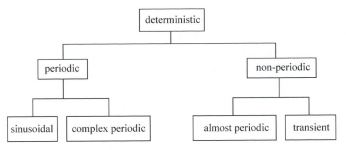

Figure 6.1. Deterministic data signals.

Complex periodic data signal

This waveform is repetitive and is reproduced exactly every time period. The time period is that associated with the fundamental frequency present ω. Fourier showed that such waveforms could be synthesized by summing a set of sine and cosine terms whose frequencies are multiples of the fundamental and that have appropriately weighted amplitudes. Multiples of a frequency are also referred to as *harmonics*, the fundamental is the first harmonic. A steady value may also need to be added to account for any offset:

$$v(t) = \frac{a_0}{2} + \sum_{n=1}^{\infty} (a_n \cos n\omega t + b_n \sin n\omega t) \tag{6.4}$$

where

$$a_n = \frac{2}{T} \int_0^T v(t) \cos n\omega t \, dt \tag{6.5}$$

and

$$b_n = \frac{2}{T} \int_0^T v(t) \sin n\omega t \, dt \tag{6.6}$$

where n is an integer $0, 1, 2, 3, \ldots$ and $T = 1/f = 2\pi/\omega$. These complex signals are the most common.

Quasi-periodic data signal

The Fourier series (equation (6.4)) shows how combinations of harmonics can produce any shape of repetitive, periodic wave. However, if three or more sinusoids are combined that are not multiples of each other then the resultant wave is not periodic. A general expression for such a waveform is

$$v(t) = \sum_{n=1}^{\infty} v_n \sin(\omega_n t + \theta_n). \tag{6.7}$$

The frequency components ω_n will not be related by rational numbers.

Transient data signal

This type of data signal is not continuous in time. Such signals often occur at the start or finish of some otherwise steady process. The relationships for some common transient responses are shown below.

$$v(t) = v_0 \exp(-at) \tag{6.8}$$

$$v(t) = v_0 \exp(-at) \cos bt. \tag{6.9}$$

These functions do not contain individual frequencies as in the previous classifications. The Fourier transform

$$V(\omega) = \int_{-\infty}^{\infty} v(t) \exp(-j\omega t) \, dt \tag{6.10}$$

illustrates that the signal contains a continuum of frequencies.

It should be remembered that whenever sampling of a steady process has a start and finish the sample record will appear to have come from a discontinuous process. Frequency components will be present in the sample record that were not in the original. This is a similar problem to aliasing described in section 5.2. To minimize the effect a long sample record should be taken.

6.2.2   Random data signals

There are many occasions when the data produced from some phenomenon is random. This may be simply because the variable being measured follows no pattern. An example of this is the signal from a temperature sensor in some environmental chamber that fluctuates because occasionally the door is opened for access. Another example of a random data signal is that of the output from a vibration sensor on an engine bearing. For a full description of these phenomena sampling would have to be continuous to infinity.

A further class of random phenomena is that of the *chaotic* system. Much interest has been generated recently in chaos theory. The mathematical relationship(s) of these systems can be written down and are often nonlinear, time-dependent, high-order, differential equations. Such equations can only be solved numerically. The criterion that classifies them as chaotic is that an infinitesimal change to the starting conditions significantly alters the outcome. In a real system this effectively means that, because of the uncertainty of the starting conditions set by the natural limits on precision, the outcome cannot be determined. Examples of chaotic systems are turbulence in fluid flow, laser output stability and the weather. To fully describe the process of a chaotic phenomenon not only would sampling have to be continuous to infinity but all other possible outcomes have to be known and recorded to infinity as well.

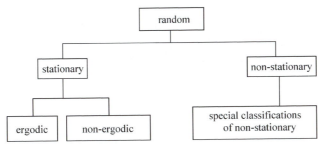

Figure 6.2. Classification of random processes.

The sub-groupings of types of random phenomena are described in figure 6.2, again following Bendat and Piersol (1986).

The storing of a data signal from a random phenomenon is called the *sample record*, $v(t)$. As mentioned above, the complete record would have to run to infinity and can then be justifiably described as the complete description of the *random process*. If a phenomenon is chaotic all possible records have to be catalogued to give a full description of the random process. Such a set of records is known as the *ensemble*, $\{v(t)\}$.

The stationary random process

Consider the collection of samples $v_1 \ldots v_n$ shown in the graphs in figure 6.3. The mean value $\bar{v}$ can be determined at some time t_1 assuming that there are an infinite number of sample records. This is said to be in the *limit* of $N \to \infty$,

$$\bar{v}(t_1) = \lim \frac{1}{N} \sum_{k=1}^{N} v_k(t_1). \qquad (6.11)$$

The mean is also-called the *first moment*.

Another statistical parameter called the *correlation* or *joint moment* can be determined using values from two different times in a sample function and then averaging over the entire ensemble:

$$R_{vv}(t_1, t_1 + \tau) = \lim \frac{1}{N} \sum_{k=1}^{N} v_k(t_1)v_k(t_1 + \tau). \qquad (6.12)$$

Because the correlation is performed on two different parts of the same signal, R_{vv} is termed the *auto-correlation* function.

Note that both $\bar{v}$ and R_{vv} are, strictly speaking, only truly representative of the random process if the whole ensemble is used, i.e. $N \to \infty$.

If the values $\bar{v}$ and R_{vv} do not change as t_1 varies, the process is said to be *stationary*. Under these conditions R_{vv} is only a function of τ. For truly stationary events all higher order moments and joint moments are also time invariant. If this is not the case the process may be referred to as *weakly* stationary; see Baher (1991).

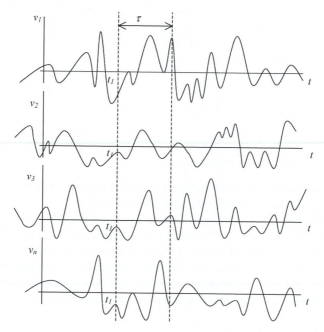

Figure 6.3. A random process ensemble of data signals.

An example of a stationary process is the noise from an engine bearing when running at some steady speed. The ensemble of sensor signals would all look very similar.

An example of a non-stationary process is the output from a pressure sensor monitoring the sequential explosions in an internal combustion engine. Each explosion would provide another set of data for the ensemble. In this case both the values of $\bar{v}$ and R_{vv} are dependent on the value of t_1.

The ergodic random process

The above calculations can be performed on a single sample record say the kth. Again, to obtain a complete description of the signal the calculations must be performed for an infinite time window, $T \to \infty$.

$$\bar{v}(k) \Rightarrow \mu_k = \lim \frac{1}{T} \int_0^T v_k(t)\, \mathrm{d}t. \tag{6.13}$$

The symbol μ is used to indicate that the mean value has been derived from a theoretical function rather than a discrete set of empirical results.

$$R_{vv}(\tau, k) = \lim \frac{1}{T} \int_0^T v_k(t)v_k(t+\tau)\, \mathrm{d}t. \tag{6.14}$$

Now, if the random process is stationary AND in addition all properties such as those derived in equations (6.13) and (6.14) are the same, whichever of the sample records is used, then the process is said to be *ergodic*. Thus,

$$\bar{v}(k) \rightarrow \bar{v} \quad \text{and} \quad R_{vv}(\tau, k) \rightarrow R_{vv}(\tau).$$

Most real life problems can be treated as ergodic. This is very fortunate since parameters for describing a random process can then be derived from a single sample function.

In the example above of an engine with vibration sensors, if every sensor was generating waveforms that produce identical values of these statistical parameters then the process is ergodic and clearly only one sensor is required. However, if the sensors were located at different places around the engine so that the waveforms were statistically different then each sensor is providing unique data about the process. The process is stationary in that the $\bar{v}$ and R_{vv} values are the same when calculated across the sample records but do not give the same values when a single record is used.

Non-stationary random processes

It is often difficult to extract any meaningful value from statistical data obtained from ensembles of these processes. Impractically long ensembles are often needed to provide sufficient data.

As an example, supposing the engine-bearing problem described above was for an aircraft. For a full description of the process the whole flight would need to be recorded since signal levels and spectra would be varying throughout the journey. However, Bendat and Piersol (1986) note that in some investigations special categories of non-stationary behaviour exist which, upon inspection, can be statistically analysed. For example, some random process of ensemble $\{v(t)\}$ could have a sample function of the form

$$v_k(t) = A(t)u_k(t) \tag{6.15}$$

where $\{u(t)\}$ is a stationary random process and $A(t)$ is a deterministic factor. If such a process can be recognized, ensemble averaging may not be necessary—a single sample record may suffice as for an ergodic process.

6.3 Characterization of random data signals

Since it is not possible to know exact time-domain descriptions due to the unpredictability of random data some form of statistical approach must be used. There are four commonly used statistical functions that can provide quantitative characterization of random data:

- mean values and variances,
- probability density function,

- auto-correlation function,
- auto-spectral density function.

6.3.1   Mean values and variances

The most commonly used statistical parameter is the average or mean. For a time-dependent signal such calculations need to be taken over the cycle. If this is not easily achievable then clearly a large number of cycles is needed so that errors caused by the inclusion of a part cycle are minimized. Calculation of the mean was described in equation (6.11). Using calculus notation,

$$\mu_v = \frac{1}{T} \int_0^T v(t)\, dt. \tag{6.16}$$

Power in any electrical or mechanical system goes as the square of the amplitude. Consequently, the mean square value may be more useful in some applications.

$$\psi_v^2 = \frac{1}{T} \int_0^T v^2(t)\, dt. \tag{6.17}$$

A further useful parameter is the variance σ^2. This determines the mean square value about the mean.

$$\sigma_v^2 = \frac{1}{T} \int_0^T (v(t) - \bar{v})^2\, dt. \tag{6.18}$$

The positive root of σ^2 is the standard deviation.

6.3.2   Probability density functions

A probability density function is useful when wishing to determine the likelihood of a particular value occurring. Consider the signal $v(t)$, shown in figure 6.4.

The probability $P(v)$ that v will lie within the range shown (see figure 6.5), v_1 and $v_1 + \Delta v$, can be estimated using the fractional time interval:

$$\frac{\Delta T_1 + \Delta T_2 + \Delta T_3 + \Delta T_4}{T} = \frac{\sum_{k=1}^N \Delta T_k}{T}. \tag{6.19}$$

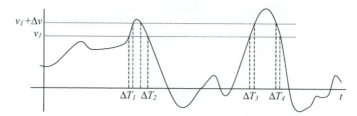

Figure 6.4. The occurrence of a waveform within some voltage limits.

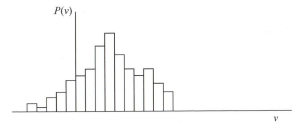

Figure 6.5. The probability of v lying within discrete intervals of Δv.

Such calculations could be performed for each interval of v across the range to give a histogram of the probabilities.

In the limit $\Delta v \Rightarrow 0$ this histogram becomes a distribution, the probability density function $p(v)$, shown in figure 6.6.

Now to calculate the probability P of v lying between v_1, $v_1 + \Delta v$ an integral is used:

$$P = \int_{v_1}^{v_1 + \Delta v} p(v) \, \mathrm{d}v. \tag{6.20}$$

Probability distributions are usually normalized meaning that the sum of all probabilities is unity:

$$\int_{-\infty}^{\infty} p(v) \, \mathrm{d}v = 1. \tag{6.21}$$

Other common statistical parameters can be calculated as follows:

$$\text{mean value} \qquad \bar{v} = \int_{-\infty}^{\infty} v p(v) \, \mathrm{d}v \tag{6.22}$$

$$\text{mean square value} \qquad \bar{v^2} = \int_{-\infty}^{\infty} v^2 p(v) \, \mathrm{d}v. \tag{6.23}$$

The calculation for the mean square value can be adapted for higher orders,

$$m_n = \int_{-\infty}^{\infty} v^n p(v) \, \mathrm{d}v \tag{6.24}$$

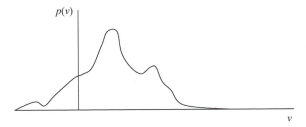

Figure 6.6. The probability density function of v.

where m_n is referred to as the nth moment. If the mean is included then m_n is called a central moment,

$$m_n = \int_{-\infty}^{\infty} (v - \bar{v})^n p(v)\, \mathrm{d}v. \tag{6.25}$$

Moments have the following properties:

1. If $p(v)$ is symmetrical in form then the 3rd, 5th, 7th etc. moments are zero. This provides a useful test of symmetry.
2. Higher order moments give greater precedence to the extremes of a signal.
3. Setting n as 2 gives the variance.

The Gaussian or normal distribution

This is by far the most commonly encountered distribution and is worth examining in more detail. It represents the distribution of some parameter x, which could be a length measurement or signal value. Gauss is reputed to have proposed this distribution when analysing astronomical data. His hypothesis was that the probability of a particular error occurring decreased with the magnitude of the error. It is found to be empirically correct. The typical form of the spread of values around the mean is shown in figure 6.7.

The general mathematical relationship describing such a bell-shaped curve is

$$p(x) = A \exp\left(\frac{-(x - \bar{x})^2}{B}\right). \tag{6.26}$$

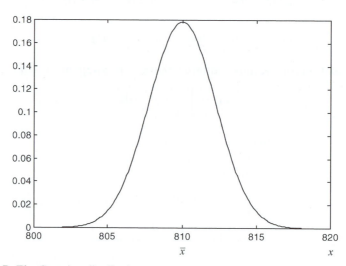

Figure 6.7. The Gaussian distribution.

Normalizing is achieved using a form of equation (6.21). Also, using equation (6.25) relates B to the variance. Combining the two results gives

$$p(x) = \frac{1}{\sigma\sqrt{2\pi}} \exp\left(\frac{-(x-\bar{x})^2}{2\sigma^2}\right) \tag{6.27}$$

which is the Gaussian or *normal* probability density function. The curve of figure 6.7 was generated using the Matlab® listing below:

```
clear, clf
mean=810; variance=5;
x= [800:.1:820]; A=1/(2*pi*variance).0.5;
B=(x-x̄).^2;
    y=A*exp(-B/(2*variance));
plot(x,y);
```

As described above in equation (6.20), to find the probability of x lying between the values a and b, an integration is performed,

$$P(x_a < x < x_b) = \int_a^b p(x)\,\mathrm{d}x \tag{6.28}$$

Fortunately, standard tables are available for evaluating such probabilities. The table value is calculated integrating from $-\infty$ to the desired value of x. Obtaining the table values for the two and then subtracting that of a from b gives the probability.

   A graph can be drawn of these tabulated $P(x)$ values versus x to give a further type of distribution known as the *cumulative* distribution function (figure 6.8).

Variance and the standard deviation

In section 6.3.1 the variance was described as the square of the standard deviation. This is not strictly true. As described in section 1.2.1, the standard deviation s is a parameter determined from a limited set of empirical values

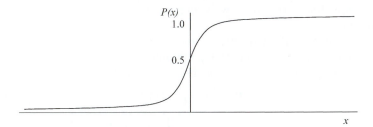

Figure 6.8. The cumulative distribution function.

not calculated using a probability distribution,

$$s = \sqrt{\frac{\sum_{i=1}^{N}(x_i - \bar{x})^2}{N}}. \tag{6.29}$$

However, in the theoretical limit for $N \Rightarrow \infty$ then $s \Rightarrow \sigma$.

Finally, it is worth noting that the point of inflection on either side of the bell-shaped curve is displaced from the mean by a value equivalent to the standard deviation. This can be illustrated by finding the turning point of the Gaussian. (The mean is taken as zero for convenience.)

$$p(x) = \frac{1}{\sigma\sqrt{2\pi}} \exp\left(\frac{-x^2}{2\sigma^2}\right) \tag{6.29}$$

$$\frac{dp}{dx} = \frac{-x}{\sigma^2} p \tag{6.30}$$

$$\frac{d^2 p}{dx^2} = \frac{-p}{\sigma^2} + \frac{x^2}{\sigma^4} p \tag{6.31}$$

which when set to zero gives $x = \pm\sigma$.

The usefulness of the standard deviation is in the estimation of the probability of an empirical value occurring at some way from the mean. For x to occur between $\bar{x}$ and

$$\pm 1\sigma, \quad P(x) = 0.683$$
$$\pm 2\sigma, \quad P(x) = 0.954$$
$$\pm 3\sigma, \quad P(x) = 0.9973.$$

Thus, the probability that a result will occur farther than $\pm 3\sigma$ from the mean is less than 2.7 per 1000.

6.3.3   Multi-variable probabilities, covariance and correlation

The use of correlation with time-dependent signals has many significant signal analysis applications. For example, it is used in echo-sounding, system identification and instrumentation. To develop the mathematical relationships for correlation it is helpful to first consider two-variable probability.

Assuming that the variables x and y are subject to probability distributions $p(x)$ and $p(y)$, then the probability of two independent events happening is simply the product of their individual probabilities,

$$P(x, y) = P(x)P(y) \tag{6.32}$$

$$P[a < x < b, c < y < d] = \int_a^b p(x)\,dx \int_c^d p(y)\,dy. \tag{6.33}$$

However, if the variables are related the probability functions are no longer separable. An example of this is the weight and height distribution of a population. The probability distribution function is now a composite $p(x, y)$. There is said to be a correlation between the two variables,

$$P[a < x < b, c < y < d] = \int_a^b \int_c^d p(x, y) \, dx \, dy. \tag{6.34}$$

Now, following the definition of the variance in equation (6.18) a co-variance C_{xy} can be determined,

$$C_{xy} = \int_{-\infty}^{\infty} \int_{-\infty}^{\infty} (x - \bar{x})(y - \bar{y}) p(x, y) \, dx \, dy. \tag{6.35}$$

If these variables are identical then equation (6.35) becomes

$$C_{xy} = \int_{-\infty}^{\infty} \int_{-\infty}^{\infty} (x - \bar{x})(x - \bar{x}) p(x) \, dx \tag{6.36}$$

so that

$$C_{xy} = \sigma_x^2 = \sigma_y^2. \tag{6.37}$$

Alternatively, if the variables are totally unrelated then

$$C_{xy} = \int_{-\infty}^{\infty} (x - \bar{x}) p(x) \, dx \int_{-\infty}^{\infty} (y - \bar{y}) p(y) \, dy \tag{6.38}$$

which yields the result

$$C_{xy} = 0 \cdot 0. \tag{6.39}$$

These results can be summarized as

$$0 \le C_{xy} \le \sigma_x \sigma_y \tag{6.40}$$

depending on the level of correlation between x and y. A correlation coefficient has been defined,

$$\rho = \frac{C_{xy}}{\sigma_x \sigma_y} \tag{6.41}$$

which lies between ± 1.

Independent random variables are uncorrelated. The converse, that uncorrelated random variables are independent, is not necessarily true but usually is in a physical system.

6.3.4 The correlation concept for time-dependent signals

As stated above the correlation of signals is an important tool in signal analysis. If x and y are signals varying in time, $x(t)$ and $y(t)$ then the correlation concept described above needs to be adapted. The changes are

1. Some suitable time window has to be chosen for the integral limits.
2. Being time-dependent signals the correlation becomes time dependent.

The two classes of correlation are *auto-correlation*,

$$R_{xx}(\tau) = \frac{1}{2T} \int_{-T}^{T} x(t)x(t+\tau)\,dt \qquad (6.42)$$

and *cross-correlation*,

$$R_{xy}(\tau) = \frac{1}{2T} \int_{-T}^{T} x(t)y(t+\tau)\,dt. \qquad (6.43)$$

The graph produced of $R(\tau)$ versus τ is called the correlelogram (auto or cross) and indicates whether the signal(s) involved, when time shifted by τ, show any correlation throughout the sampling window $2T$. The function can be easily determined with an algorithm,

$$R_{xx}(r\,\Delta t) = \frac{1}{N} \sum_{j=1}^{N} x_j x_{j+r} \qquad (6.44)$$

where $r\,\Delta t$ represents the discrete τ values, Δt being the sampling period and r an integer beginning at zero.

   If a correlation peak does occur at some value of τ a similar expression to equation (6.41) can be used to estimate the level of correlation.

6.3.5   Auto-correlation and spectral density functions

Power spectra and the correlation function

It is often useful in engineering to know how the power in a signal or an oscillating system is distributed throughout the frequency spectrum. The *power spectral density function G* indicates the power distribution in a signal and has units of watts per radian or watts per hertz. The mean power $\bar{W}$ within some bandwidth can be obtained from G,

$$\bar{W} = \int_{f_1}^{f_2} G\,df. \qquad (6.45)$$

Now, the auto-correlation of a voltage signal $v(t)$ has units of volts^2 which is the power dissipated in a unit resistor. Intuitively, one might think that there ought to be some relationship between the auto-correlation and the power spectral density function. To determine this relationship it is necessary to begin with the Fourier transform of the auto-correlation. For some signal of voltage $v(t)$ equation (6.42) is written as

$$R_{vv}(\tau) = \frac{1}{2T} \int_{-T}^{T} v(t)v(t+\tau)\, \mathrm{d}t. \tag{6.46}$$

Because R_{vv} is a function of τ, '$\exp(-\mathrm{j}\omega\tau)\, \mathrm{d}\tau$' is used to perform the Fourier transform of equation (6.10),

$$\Im[R_{vv}(\tau, T)] = \int_{-\infty}^{\infty} \left[\frac{1}{2T} \int_{-T}^{T} v(t)v(t+\tau)\, \mathrm{d}t \right] \exp(-\mathrm{j}\omega\tau)\, \mathrm{d}\tau. \tag{6.47}$$

In order to align the limits of integration a large sampling interval is needed. Making $T \to \infty$ gives

$$\Im[R_{vv}(\tau, T)] = \frac{1}{2T} \int_{-\infty}^{\infty} \int_{-\infty}^{\infty} v(t)v(t+\tau)\exp(-\mathrm{j}\omega\tau)\, \mathrm{d}t\, \mathrm{d}\tau. \tag{6.47}$$

Re-arranging,

$$\Im[R_{vv}(\tau, T)] = \frac{1}{2T} \int_{-\infty}^{\infty} \left[\int_{-\infty}^{\infty} v(t)\exp(\mathrm{j}\omega t)\, \mathrm{d}t \right] v(t+\tau)\exp(-\mathrm{j}\omega(t+\tau))\, \mathrm{d}\tau \tag{6.48}$$

which are effectively two Fourier transforms. Writing $V(-\mathrm{j}\omega)$ as the transform of $v(t)$ in the square brackets above gives

$$\Im[R_{vv}(\tau, T)] = \frac{1}{2T} \int_{-\infty}^{\infty} V(-\mathrm{j}\omega)v(t+\tau)\exp(-\mathrm{j}\omega(t+\tau))\, \mathrm{d}\tau. \tag{6.49}$$

Substituting $\lambda = t + \tau$,

$$\Im[R_{vv}(\tau, T)] = \frac{1}{2T} V(-\mathrm{j}\omega) \int_{-\infty}^{\infty} v(\lambda)\exp(-\mathrm{j}\omega\lambda)\, \mathrm{d}\lambda \tag{6.50}$$

revealing another Fourier transform. Thus,

$$\Im[R_{vv}(\tau, T)] = \frac{1}{2T} V(-\mathrm{j}\omega)V(\mathrm{j}\omega) \tag{6.51}$$

which can be written as

$$\Im[R_{vv}(\tau, T)] = \frac{1}{2T} |V(\mathrm{j}\omega)|^2. \tag{6.52}$$

It is not immediately obvious what equation (6.52) means. To interpret this relationship it is useful to begin with an expression for the total energy dissipated in a unit resistor during the time interval $-T$ to T. For a signal of voltage $v(t)$, the energy dissipated during the time window $2T$ would normally be estimated using

$$E = \int_{-T}^{T} v^2(t)\, \mathrm{d}t. \tag{6.53}$$

But another expression for E can be obtained using Parseval's theorem which states that

> the total energy in a signal is the sum of the energies in each of its frequency components.

Again for a unit resistor this can be expressed as

$$E = \int_{-\infty}^{\infty} |V(j\omega)|^2 \, d\omega. \tag{6.54}$$

Dividing equation (6.54) by $2T$ gives the average power

$$\bar{W} = \frac{1}{2T} \int_{-\infty}^{\infty} |V(j\omega)|^2 \, d\omega. \tag{6.55}$$

However, equation (6.55) is the integral of equation (6.52). These can be equated if equation (6.55) is first differentiated with respect to ω,

$$\frac{d\bar{W}}{d\omega} = \frac{1}{2T} |V(j\omega)|^2 = S_{vv}(\omega) \tag{6.56}$$

where $d\bar{W}/d\omega$ is the power spectral density function labelled here as $S_{vv}(\omega)$. Comparing equation (6.56) with equation (6.52) gives

$$\Im[R_{vv}(\tau, T)] = S_{vv}(\omega). \tag{6.57}$$

So, the Fourier transform of the auto-correlation function produces the power spectral density function. For this reason $S_{vv}(\omega)$ is also-called the *auto-spectral density* function. Auto-spectral density is the fourth of the descriptive techniques listed in section 6.3.

The Fourier transform of the auto-correlation shown above and its inverse are known as the Wiener–Khintchine relations. The general expression of the transform is

$$S_{xx}(\omega) = \int_{-\infty}^{\infty} R_{xx}(\tau) \exp(-j\omega\tau) \, d\tau \tag{6.58}$$

which can be re-written as

$$S_{xx}(\omega) = \int_{-\infty}^{\infty} R_{xx}(\tau) \cos\omega\tau \, d\tau. \tag{6.59}$$

The $-j\sin\omega t$ term disappears because S_{xx} and R_{xx} are real functions, not complex. Also, since the cosine function is symmetrical the integral can be limited to positive values of τ,

$$S_{xx}(\omega) = 2 \int_{0}^{\infty} R_{xx}(\tau) \cos\omega\tau \, d\tau. \tag{6.60}$$

The inverse relationship is

$$R_{xx}(\tau) = \frac{1}{2\pi} \int_{-\infty}^{\infty} S_{xx}(\omega) \exp(j\omega t)\, d\omega \qquad (6.61)$$

or

$$R_{xx}(\tau) = \frac{1}{\pi} \int_{0}^{\infty} S_{xx}(\omega) \cos \omega\tau\, d\omega. \qquad (6.62)$$

Weiner (USA) and Khintchine (USSR) independently derived these relationships in the early 1930s.

The practical power spectral density function

The careful reader will have noticed that two separate symbols have been used in this section to represent power spectral density, G and S_{vv}. This was an attempt to distinguish practical usage from theoretical derivation. The function $S_{vv}(\omega)$ came from a Fourier transform and because the signal was taken as continuous throughout time ranging from $-\infty$ to ∞ the resulting frequency range has the same limits. Negative frequencies, however, are only of theoretical use when dealing with integrals. For practical use the *single-sided* power spectral density function G is employed. Now, since $R_{vv}(\omega)$ is determined by multiplying $v(t)$ with itself it must be an even function, meaning symmetrical around the ordinate axis,

$$R_{vv}(\tau) = R_{vv}(-\tau). \qquad (6.63)$$

It follows that $S_{vv}(\omega)$ must be the same. Consequently, $G(\omega)$ can be obtained from $S_{vv}(\omega)$ using only positive frequencies, so

$$G(\omega) = 2S_{vv}(\omega) = 2\Im[R_{vv}(\tau, T)] \qquad \text{for } 0 \leq \omega \leq \infty. \qquad (6.64)$$

Also, from equation (6.56)

$$G(\omega) = \frac{1}{T}|V(j\omega)|^2 \qquad \text{for } 0 \leq \omega \leq \infty. \qquad (6.65)$$

This can be a convenient method of performing a spectral analysis on a time-domain signal. The auto-correlation and Fourier transform functions of equations (6.64) and (6.65) can be efficiently computed using algorithms in a dsp microprocessor system. See Brook and Wynne (1988) for a fuller description. As these authors point out, historically equation (6.64) was used to determine frequency spectra. This was because the discrete Fourier transform (DFT) is computationally intense requiring $4(N)^2$ multiplications and $2(N)^2$ additions where N is the number of spectral coefficients. However, the publication of the fast Fourier transform (FFT) by Cooley and Tukey (1965) meant that equation (6.65) can be computed directly. Interestingly, because of the performance of the FFT some signal analysers now determine the auto-correlation function by first performing the FFT on a signal to obtain $S_{vv}(\omega)$ and then, from equation (6.61), obtain $R_{vv}(\tau)$ using the inverse FFT.

6.4   Noise

6.4.1   Origins of noise

As explained in section 6.1, noise is a random signal and is treated statistically. There are three processes that generate noise.

(a) Thermal noise

The thermal motion of charge carriers in any circuitry produces tiny random currents at the output terminals. This random motion has a mean of zero. However, both the r.m.s. current and voltage are not zero and become noticeable when the signal being recovered is weak. Thermal noise is also sometimes referred to as Johnson or Nyquist noise. The integrated mean-square-voltage in the frequency band Δf is given by

$$\Delta \bar{v}^2 = 4kTR\,\Delta f \qquad (6.66)$$

where k is Boltzmann's constant, T the absolute temperature and R the resistance of the noise source. The spectrum is flat and so, by analogy to white light, is sometimes referred to as *white* noise. Equation (6.1) can be re-written as a *power spectral density function,*

$$\frac{1}{R}\frac{\Delta \bar{v}^2}{\Delta f} = 4kT \quad \text{[watts per hertz].} \qquad (6.67)$$

The power spectral density for thermal noise is shown in figure 6.9.

(b) Shot noise

This form of noise is produced by the random arrival of charge carriers and becomes noticeable when current flow is tiny. It can be shown using a Poisson

Figure 6.9. Power spectral density of thermal noise.

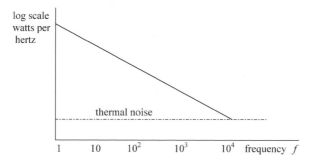

Figure 6.10. Power spectral density of flicker noise.

statistical analysis that the integrated mean-square-current is given by

$$\Delta \bar{i}^2 = 2ei_0 \, \Delta f \qquad (6.68)$$

where e is the charge on the electron and i_0 is the mean current. The power spectral density function is spectrally flat, similar to figure 6.9.

(c) Flicker noise

A further type of noise occurs in semiconductors and some composition materials. The power distribution declines with frequency and so is referred to as *one-over-f* or *flicker* noise in semiconductors (Nicollian and Brews 1982). Germanium devices tend to be worse than silicon. Flicker noise does not occur in metallic components. The log of the power spectral density is shown in figure 6.10. The integrated mean-square-voltage is given by

$$\Delta \bar{v}^2 = \frac{Bi^2}{f} \, \Delta f \qquad (6.69)$$

where B is an empirical constant. It is more dominant than thermal noise below 10 kHz. Flicker noise occurs in some composition resistors and so, when dealing with low-level signals having frequencies of less than 10 kHz, metallic thin-film or wire-wound resistors should be used.

Amplification

When amplifying low-level signals, amplifiers having low noise levels in their input stages must be sought. Care must be taken when mounting these, particularly when high input impedance devices are needed. A large number of low-noise amplifiers are commercially available and manufacturers sometimes include recommendations on circuit layout.

However, the measurement of a weak, steady variable poses further problems to the signal conditioning. This is because the intermediate

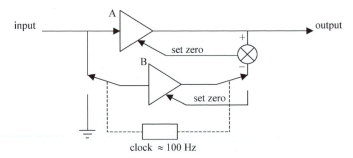

Figure 6.11. Schematic diagram of an auto-zeroing amplifier.

stages of the system have to be directly coupled and any amplifier zero error or drift will contribute to the measured value.

The commutating, auto-zeroing amplifier may be a suitable choice to reduce the effects of drift. This is a development following on from the chopper-stabilized amplifier that converts a d.c. signal to a.c. then uses a.c. amplification and subsequently returns the signal to d.c. The advantage of the commutating, auto-zeroing amplifier is that it is constantly on-line. There are a number of commercially available versions of this CMOS integrated circuit package. As illustrated in figure 6.11, this device constantly checks and corrects its own output drift. The input to the subsidiary amplifier B is grounded and its output then used to reset its own zero. At the clock signal the output of the identical main amplifier A is subtracted from the output of B. Any discrepancy is then used to reset the zero of A. Note that amplifier A is always on-line. This helps to reduce clock-induced ripple in the output. Zero stability is a factor of five better than precision, bipolar-technology, operational amplifiers. The disadvantages of the device are

- external capacitors are required for storing the zero-error correction voltage,
- common mode voltage is restricted, $+3.5\,\text{V}$ to $-5\,\text{V}$,
- current output is limited, $5\,\text{mA}$,
- if output saturation occurs recovery can be slow,
- supply voltage V^+ to V^- is limited to $18\,\text{V}$.

6.4.2   Noise reduction techniques

Whenever such noise phenomena occur a number of options are available to enhance the signal-to-noise ratio.

Filtering

The noise spectra shown above are extended in frequency. If the signal and noise are measured directly then a large contribution to the resultant value

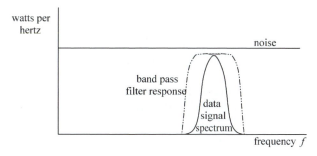

Figure 6.12. Noise filtering.

will come from the noise. A simple filter, placed in the signal conditioning circuitry input, can be very effective if the signal is known to be in a restricted part of the spectrum. Depending on the position of the data signal spectrum low, high or band pass filters are used. See figure 6.12.

Simple time-averaging

Simple averaging can be used if the variable is static or slowly changing with time, such as might occur with the temperature of a liquid in a large tank. A low pass filter as described above is, in-effect, an analogue averaging device.

   With the advent of cheap memory, averaging can be done more sophisticatedly using digital techniques. For most types of noise the fluctuation of the amplitude of the current or voltage of the signal with time $x(t)$ is a Gaussian distribution around the mean. The determination of the mean value of x and its standard deviation were described in section 6.3.1 from equations (6.16) and (6.18). For practical use the discrete versions of these functions are employed,

$$\bar{x} = \frac{\sum_{i=1}^{N} x_i}{N} \tag{6.70}$$

where the number of samples N taken within some time window T might be typically 1024. The standard deviation of the samples is a measure of the spread of the signal around the mean and represents the level of noise. When $\bar{x}$ is zero, as is the case for pure noise, s is the root-mean-square of x. It is determined using

$$s = \sqrt{\frac{\sum_{i=1}^{N} x_i^2}{N}}. \tag{6.71}$$

As described in section 1.2.1, to gain greater confidence in the value of $\bar{x}$ a number of windows are sampled and the mean $\bar{x}_m$ of all the $\bar{x}$ values determined. It can be shown that the standard deviation s_m for this ensemble of mean values taken from p windows is determined using

$$s_m = \frac{s}{\sqrt{p}} = \sqrt{\frac{\sum_{i=1}^{N} x_i^2}{pN}}. \tag{6.72}$$

Clearly, increasing N or p reduces s_m (the spread of the distribution) and so a better estimate of the measurand is achieved. For example, increasing N or p by a factor of 10 improves the signal-to-noise ratio by $-10\,dB$. But obviously the total sampling time pT needs to be short compared with the time for any change in the variable to occur.

A refinement to this technique is to perform a *rolling average*. Here the mean value from the last window to be sampled is averaged with the stored mean to give a new stored mean. This of course has the effect of awarding a greater significance to the last window sampled. Variations of the rolling average technique have been developed where the mean values from the p windows to be averaged are all weighted with progressively less significance the further back in time that they occurred. Various weighting functions have been used.

These simple averaging procedures cannot be used when the noise distribution is asymmetrical about zero since this would give a non-zero value for $\bar{x}$ which would be incorporated with the measurand.

Sophisticated time-averaging

As indicated above there are limitations on the ultimate performance of the simple averaging process. Firstly, the variable to be measured must be virtually stationary and secondly there is the increase in $1/f$ noise at low frequencies. The following techniques are more sophisticated averaging processes that require that the signal is repetitive. This can often be arranged either by driving the sensor with some form of periodic excitation or by arranging that the measurand occurs repetitively. Such techniques are effectively placing the data signal in a different part of the frequency spectrum. A further advantage is that any amplification can be a.c. coupled reducing the problems of drift.

The multi-channel scaler

For periodic or repetitive transient signals averaging can be performed with a *multi-channel scaler*. This digital device consists of a number of channels, 1024 for example, which store the signal. If the repetition period is T a time resolution Δt of $T/1024$ is achieved. Precision triggering is essential so as to sample at the same point in the waveform at every scan. Up to 10^6 scans can be stored and averaged. Commercial systems often employ a counting technique. Samples of the waveform are fed into a voltage-to-frequency converter. This frequency is counted for the time Δt. The count

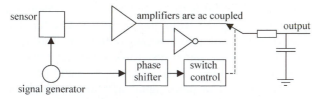

Figure 6.13. Phase-sensitive detection.

is then stored in the appropriate channel. Counts from subsequent scans are either added or subtracted depending on the polarity of the sample. A channel acts like the accumulator of an up–down counter. After a large number of scans the noise tends to average towards zero, as does any unwanted repetitive signal having a period different from T. The bandwidth is approximately equal to the inverse of the total sampling time.

The phase-sensitive detector

Phase-sensitive detection is a generic technique that can be implemented in a number of forms. In all cases phase adjustment between a reference signal and the data signal is available. Early commercial designs were entirely analogue. Digital hybrids provide better stability. The basic principle is illustrated in figure 6.13. A deflection bridge circuit driven sinusoidally, such as might be the case for a strain gauge system, provides a useful example; see Wilmshurst (1990). The phase-shifter is used to compensate for any system-induced phase-shift. The output-smoothing filter performs an averaging function. It has the effect of reducing the system bandwidth. It must pass slowly varying data signals but reject the harmonics generated.

Frequency domain analysis is a useful approach in revealing the relative merits of the different options when using phase-sensitive detection. To conduct the analysis an analogue multiplier is substituted in place of the switch. This provides a general model of the phase-sensitive detector, as illustrated in the schematic of figure 6.14, enabling the performance of different experimental configurations to be modelled.

Three multiplication types are described below with an appropriate application to illustrate the concept.

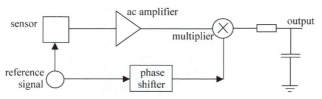

Figure 6.14. Generalized phase-sensitive detector.

1. *Bipolar square wave.*   This is the equivalent of the system in figure 6.13, the example given being the strain gauge bridge. The bipolar waveform performs the function of the inverting amplifier. For the equivalent performance the reference signal generator must have two outputs, one sinusoidal the other bipolar. This square-wave action has the effect of producing harmonics. The data signal appears as sidebands at these harmonics, which also have the effect of increasing the noise.

2. *Unipolar square wave.*   In some optical experiments, in order to use the phase-sensitive detection technique, the beam of light is interrupted by a chopper wheel rotating at high speed instead of having a signal generator. The phenomenon under investigation is made to occur at a frequency set by the light pulse repetition rate. A reference signal is taken from the wheel using a separate photodiode. The waveform of the data signal is of a unipolar format with data + noise in one half cycle and noise alone in the other. There is a significant disadvantage to the unipolar waveform over that of the bipolar. It has the same harmonics but, in addition, a d.c. component. This results in the system accepting off-set drift and $1/f$ noise. The reason for using the unipolar format in an optical experiment is the convenience of modulating the light beam using the chopper wheel. A more general advantage in other applications is that analogue multipliers have an upper frequency limit of order MHz whereas switching circuits can operate at GHz.

3. *Sinusoidal waveform.*   The disadvantages described above when using square waveforms do not occur when two pure sinusoids are multiplied together. This is because no harmonics are produced. The multiplication of two sinusoids results in two new sinusoids: one whose frequency is the sum of the original frequencies and one whose frequency is the difference (equation (6.81)). If the two original frequencies are identical the result is then a single sinusoid at twice the original frequency. This technique has been successfully applied in spectroscopy. A range of wavelengths is used to irradiate the material being investigated. At a particular wavelength some peak (or trough) will occur. This is illustrated in figure 6.15(i). To locate this peak the wavelength is increased steadily during the experiment but, in addition, it is made to oscillate with small amplitude throughout the much longer sweep cycle (figure 6.15(ii)). The result is to generate a data signal modulated at the oscillating frequency the amplitude of which depends on the steepness of the edge of the peak (figure 6.15(iii)). Also, a phase-shift occurs between the waveforms on the up and down sides of the peak. Following multiplication with the oscillation signal a waveform is produced (figure 6.15(iv)) having an asymmetry centred on the peak cross-over. With a smoothing filter the peak cross-over is clearly seen in figure 6.15(v). This waveform is effectively the differential of the spectroscopy peak of figure 6.15(i). For the system to function correctly the time constant of the filter must be long compared with the

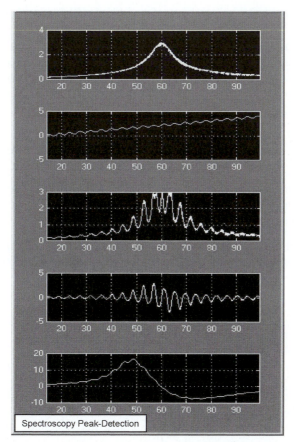

Figure 6.15. Spectroscopy using phase-insensitive detection. Top to bottom: (i) a transmission peak (shown scanned over the same time as the other results for comparison), (ii) the scanning signal, (iii) the output of the optical sensor, (iv) the output of the multiplier block, (v) the smoothed output.

time period of the oscillations but short compared with the sweep time. If the sweep time is too fast for the smoothing circuit the cross-over point will shift.

Generally, whatever waveform is used in a phase-sensitive detector the two half-periods must be of exactly the same duration otherwise the level of the recovered signal will be reduced.

The lock-in amplifier

Phase-sensitive detectors can be built from discrete elements but it is usually more cost-effective to use commercially available systems, which are described as lock-in amplifiers. See Meade (1983) for a detailed description.

Commercial devices contain a variable-frequency signal generator with tracking filter, a low-noise wide dynamic range amplifier, a high performance linear phase detector and a post-detection filter with a choice of time constants. Provision is made for the use of an external signal generator. Phase adjustment is done manually by the operator in order to get the best signal.

Auto-correlation

One powerful application of correlation is to the recovery of signal information using auto-correlation. Auto-correlation was described earlier in sections 6.2.2 and 6.3.4. By correlating a signal with itself it is possible to

(i) derive some statistical information from a random signal,
(ii) separate sinusoids from a random signal,
(iii) determine the period of non-sinusoidal repetitive waveforms.

The statistical application was described in section 6.3. The following paragraphs show how a sinusoid can be recovered when buried in noise and this same concept can also be used to determine the period of a non-sinusoidal waveform.

Auto-correlation of the sum of two signals

Supposing the signal $z(t)$ consists of two parts,

$$z(t) = x(t) + y(t) \tag{6.73}$$

where $x(t)$ is the desired data and $y(t)$ is noise. Then, from equation (6.14), reproduced here, an auto-correlation is determined,

$$R_{vv}(\tau) = \frac{1}{2T} \int_{-T}^{T} v(t)v(t + \tau) \, \mathrm{d}t. \tag{6.74}$$

So the auto-correlation of equation (6.73) can be expressed as

$$R_{zz} = \frac{1}{2T} \int_{-T}^{T} [x(t) + y(t)][x(t + \tau) + y(t + \tau)] \, \mathrm{d}\tau \tag{6.75}$$

$$R_{zz} = \frac{1}{2T} \int_{-T}^{T} [x(t)x(t + \tau) + x(t)y(t + \tau) + y(t)x(t + \tau) + y(t)y(t + \tau)] \, \mathrm{d}\tau$$

$$\tag{6.76}$$

$$R_{zz} = R_{xx} + R_{xy} + R_{yx} + R_{yy}. \tag{6.77}$$

The terms R_{xy}, R_{yx} are cross-correlations terms of the correlation between x, y. If there is no correlation then these two are both zero. If the unwanted signal is true random noise then there will be no correlation between noise and data. So

$$R_{zz} = R_{xx} + R_{yy}. \tag{6.78}$$

Auto-correlation of a sine wave

The data signal $x(t)$ can be expressed as

$$x_k(t) = x_0 \sin[\omega_0 t + \theta(k)]. \tag{6.79}$$

The subscript k is used to indicate that the data could be part of some ensemble $\{x(t)\}$ with $\theta(k)$ being some unbiased random phase-shift. The auto-correlation function can be performed across the ensemble if the data is ergodic. The auto-correlation function then becomes

$$R_{xx}(\tau) = \frac{x_0^2}{2\pi} \int_0^{2\pi} \sin(\omega_0 t + \theta) \sin[\omega_0(t + \tau) + \theta] \, d\theta. \tag{6.80}$$

Using the trigonometrical identity,

$$\sin A \sin B = \cos(A - B) - \cos(A + B) \tag{6.81}$$

gives

$$R_{xx}(\tau) = \frac{x_0^2}{2} \cos \omega_0 \tau. \tag{6.82}$$

Performing the auto-correlation across the ensemble of sample records, rather than along a single sample record, gives the important result that the auto-correlation function of the kth sample is independent of any phase shift present and is the same frequency.

Auto-correlation of noise

The auto-correlation of noise is not calculated in the way used above. This is because the time-dependent voltage signal is random and so no analytical expression can be substituted into the integral for $R(\tau)$. Instead the value for the auto-correlation is determined using the Wiener–Khintchine transformation, which converts the power spectral density function $S(f)$ into the auto-correlation directly,

$$R_{xx}(\tau) = \int_{-\infty}^{\infty} S_{xx}(f) \exp(j\pi\tau f) \, df. \tag{6.83}$$

In practice the frequency range is positive only so, writing $G(f)$ for the real power spectral density, equation (6.83) becomes

$$R_{xx}(\tau) = \int_0^{\infty} G_{xx}(f) \cos 2\pi f \tau \, df. \tag{6.84}$$

Note that the complex sine term is ignored since, as stated in a previous section, auto-correlations are always even functions of τ.

For a true white noise spectrum all frequencies are represented with equal amplitude so $G_{xx}(f) = a$. Substituting in equation (6.83) with

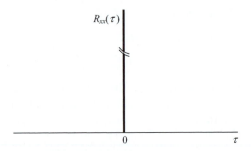

Figure 6.16. The auto-correlation of pure white noise.

$S_{xx}(f) = \frac{1}{2}a$ gives

$$R_{xx}(\tau) = \frac{a}{2} \int_{-\infty}^{\infty} \exp(\mathrm{j}2\pi f\tau)\,\mathrm{d}f. \tag{6.85}$$

But this can be evaluated easily by referring to the Fourier transform pair for a delta function, $\delta(t)$. A delta function is an infinitely tall spike at the origin and infinitesimally narrow such that the area is unity:

$$\Im[\delta(t)] = \int_{-\infty}^{\infty} \delta(t)\exp(\mathrm{j}2\pi ft)\,\mathrm{d}t = 1. \tag{6.86}$$

But the inverse Fourier transform of unity is

$$\Im^{-1}[1] = \int_{-\infty}^{\infty} \exp(\mathrm{j}2\pi ft)\,\mathrm{d}f = \delta(t). \tag{6.87}$$

Thus by comparing expressions it can be seen that

$$R_{xx}(\tau) = \frac{a}{2}\delta(\tau). \tag{6.88}$$

Figure 6.16 is a sketch of this origin-centred delta function. Noise in practice will not extend to infinity due to the bandwidth limit of the system. So, to evaluate equation (6.84) for a real system bandwidth limited noise is used. The power spectral density function is shown in figure 6.17.

Substituting this function into equation (6.84) gives

$$R_{xx}(\tau) = \int_{f_0 - 1/2B}^{f_0 + 1/2B} a\cos(2\pi f\tau)\,\mathrm{d}f \tag{6.89}$$

$$= a\left[\frac{\sin 2\pi f\tau}{2\pi\tau}\right]_{f_0 - 1/2B}^{f_0 + 1/2B} \tag{6.90}$$

$$= \frac{a}{\pi\tau}(\sin\pi\tau B)\cos 2\pi f_0\tau \tag{6.91}$$

$$= aB\left[\frac{\sin\pi\tau B}{\pi\tau B}\right]\cos 2\pi f_0\tau. \tag{6.92}$$

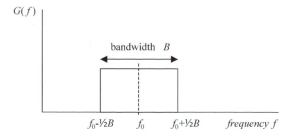

Figure 6.17. Power spectral density function of bandwidth limited noise.

For the special case where the frequency bandwidth ranges from 0 to B, so that $f_0 = \frac{1}{2}B$, then using the trigonometrical identity,

$$\sin \theta \cos \theta = \tfrac{1}{2}\sin 2\theta \qquad (8.93)$$

gives

$$R_{xx}(\tau) = aB\left(\frac{\sin 2\pi\tau B}{2\pi\tau B}\right). \qquad (8.94)$$

Functions of the type $(\sin x)/x$ are sometimes called sinc functions. The form of the function is shown in figure 6.18.

Finally, it can be seen that if a combined signal of a sine wave buried in noise is auto-correlated then the autocorrelogram will separate the two. The separation will increase the further from the origin.

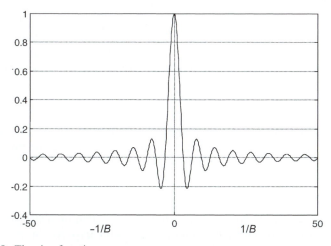

Figure 6.18. The sinc function.

6.5   Interference

6.5.1   Sources of interference and coupling mechanisms

As explained in section 6.1, interference is a deterministic phenomenon. Its origins are virtually all from electromagnetic waves produced by

- local power cables either external to the system or badly routed within it,
- d.c. circuits which are being switched and consequently generating electro-magnetic transients,
- electrical discharges from high voltage terminals (a.c. lines produce inter-ference but corona discharge from high voltage d.c. will generate noise),
- arcing in electrical machinery,
- fluorescent lighting generating waves at twice the power line frequency,
- digital circuitry,
- radio transmissions,
- microwave equipment.

Inductive coupling

From Faraday's law, the induced emf in some secondary circuit element can be expressed in terms of the current in the primary element,

$$\text{emf} = M \frac{dI_p}{dt}. \tag{6.95}$$

The value of M is set by the geometry of the system. The general expression for M was first described by Neuman; see Bleaney and Bleaney (1965):

$$M = \frac{\mu\mu_0}{4\pi} \iint \frac{ds_1 \cdot ds_2}{r} \tag{6.96}$$

where ds_1 and ds_2 are the current elements and r the separation. Its analytical solution is difficult in all but the simplest geometrical arrangements. For example, two co-axially mounted plane coils whose separation z is large compared with their radii a and b and having n_1 and n_2 turns respectively is

$$M = \frac{\mu\mu_0 \pi a^2 b^2 n_1 n_2}{2z^3}. \tag{6.97}$$

A common occurrence of this coupling type is that of the magnetic field produced near a power transformer being coupled into some pre-amplified circuitry containing weak signals.

Capacitive coupling

Alternating electric fields can induce charge movement on nearby conducting objects effectively forming a capacitance across the gap. Again, the analytical determination for the value of the capacitance can be difficult to derive in all

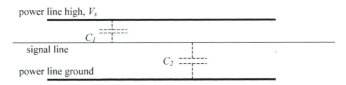

Figure 6.19. Coupling by stray capacitance.

but the simplest cases. The simplest case is that of the two parallel plates,

$$C = \frac{\varepsilon_0 \varepsilon_r A}{d} \qquad (6.98)$$

where d is the plate separation and A is their common area of overlap. Another useful expression is that of the capacitance per unit length between two parallel, long wires,

$$C = \frac{\pi \varepsilon_0 \varepsilon_r}{\log_e(d/a)} \quad [\mathrm{F\,m^{-1}}] \qquad (6.99)$$

where d is the separation of their centres and a their radii. When a wire is close to and running parallel with a large conducting plane a similar expression is obtained,

$$C = \frac{2\pi \varepsilon_0 \varepsilon_r}{\log_e(2h/a)} \quad [\mathrm{F\,m^{-1}}] \qquad (6.100)$$

where h is the separation. A simple example of how coupling might occur is shown in figure 6.19.

The stray capacitances form a potential divider across the supply lines. The signal line is effectively tapping into the electric field across the gap giving an additional signal,

$$\frac{X_2}{X_1 + X_2} V_s = \frac{1/\mathrm{j}\omega C_1}{1/\mathrm{j}\omega C_1 + 1/\mathrm{j}\omega C_2} V_s = \frac{C_2}{C_1 + C_2} V_s. \qquad (6.101)$$

Loop circuits

Closed loop circuits can be inadvertently created by connecting leads, on printed circuit boards or with earth connections as shown later in figure 6.21. Such loops can act as antennae and consequently introduce electromagnetic waves into the system.

The following two phenomena are not strictly coupling mechanisms but rather injection processes.

Multiple earth connections

It is useful to first consider some terminology. Many use the words *earth* and *ground* synonymously. Earth is more commonly used in the United Kingdom

(a)   (b)

Figure 6.20. Earth-ground symbols: (a) earth or ground, (b) chassis or frame.

than in North America. There are even two different circuit symbols (figure 6.20) employed, one for earth or ground the other for the equipment chassis or frame. The British Standard is BS 3939, EN 60617-2.

Ott (1988) gives a standard definition of a ground as 'an equipotential point or plane that serves as a reference potential for a circuit or system', in that 'its voltage does not change regardless of the current applied to it or drawn from it'. He goes on to point out that this is not achievable in practice and suggests that a better definition might be 'a low impedance path for current to return to source'.

There are two purposes for grounds: for safety and for the signal reference. Neither the real earth, earth cabling nor ground-planes have zero potential difference throughout. The resistance of soil can vary by several orders of magnitude due to variation in composition and water content; see Kodali (1996). The value of 1 Ω per metre length could be taken as a nominal order of magnitude estimate. By contrast, copper cable of 2 mm diameter has a resistance of about 5 mΩ per metre. Consequently, if a current is passing through the earth or down an earthing cable a potential difference will exist between any sample points. Heavy electrical machines and faulty equipment whose fault is below the trip level may cause earth leakage. Also, stray magnetic fields may cause eddy currents by inductive coupling into the earth.

These faults will have the effect of injecting an extra potential difference in any signal lines which are using earth cabling as part of the circuitry, such as with ordinary co-axial cable, or as part of some bonding safety system (figure 6.21).

Such leakage currents are at the power frequency. These *earth loops* can be responsible for *hum* in studio audio systems. An interesting example is that of the electric locomotive or tram that uses the track as the circuit return.

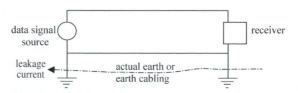

Figure 6.21. Injected interference from multiple earth connections.

Without the use of a Faraday cage shrouding the track foundation, tens of volts may occur in the soil near the track as the vehicle passes.

It is also possible for an earth loop to act as an antenna in the way described above under *loop circuits*. It is the higher frequencies that cause interference via this mechanism rather than the power line frequency.

Printed circuit boards

A result similar to that caused by multiple earthing described above may occur with the ground plane on a printed circuit board. The reason, however, is that the signal and/or power line ground may not remain everywhere at zero volts if the currents being returned along the track are too large for the size of copper strip laid down. Any potential difference occurring between different parts of the ground plane will effectively introduce a signal into the system.

It is a common experience of circuit board designers that signals produced in some section of a system are carried through the supply rails into other sections. This is a particular problem at high frequencies. It is as a result of the supply rails not being able to deliver the power required by a section so that the supply voltage is no longer steady but ripples at the oscillating or switching frequency. This has the effect of causing other sections of the system to be injected with the same frequency. CMOS components are generally more prone to this problem than bipolar.

Co-axial cable

Early co-axial cable suffered from being sensitive to movement; see Perls (1952). Bending of the cable causes friction between layers that can create charge on the layer surfaces. If, in the bending process, contact is lost between some small part of the insulation and the conductor the charges are not conducted away. This results in an additional emf at the terminals. The initial solution was to coat the insulation in graphite powder so as to maintain electrical continuity. However, modern manufacturing techniques have enabled better physical bonding to occur between conductors and insulation.

It is still good practice not to allow co-axial cable free movement when carrying tiny signals.

6.5.2 Interference reduction techniques

Geometrical isolation

The inductive and capacitive mechanisms described above are significantly dependent on the physical parameters of the layout. Simply increasing the physical separation of the components will decrease the coupling. The interference will decrease with the inverse of the distance or better.

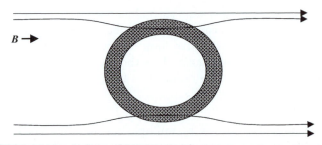

Figure 6.22. Magnetic shielding.

Shielding

The exclusion of steady or low frequency magnetic and electric fields can be achieved by enclosing the sensitive circuitry within metallic shields.

Using Gauss's theorem it can be shown that electrostatic charge must reside on the outside of a conductor. Steady electric fields cannot be sustained within metals since any potential gradient would cause charge to move until the field is zero. Consequently, the outer surface becomes an equipotential where any induced or transferred charge resides. These concepts are also true for a hollow metallic conductor. This means that a metallic box is an effective shield to steady, external electric fields. It is possible to use a metallic mesh for shielding provided that the holes are small compared with the distance between the screen and circuitry to be protected.

The flow of magnetic flux depends on the reluctance of the flux path. The reluctance for a particular material specimen goes as

$$\Re = \frac{l}{\mu_0 \mu_r A} \tag{6.102}$$

where A is the cross-sectional area and l the path length within the material. The reluctance is lowest in a path containing material of high relative permeability μ_r. If a closed container is constructed out of such a material then the magnetic field lines will remain within this and so shield the internal circuitry (figure 6.22).

A commonly used material for the shield is the alloy *mu-metal*. It is often employed for the construction of shields for cathode ray tubes.

Whenever a shielding enclosure is used it is good practice to maintain a significant gap between circuitry and enclosure. This helps to ensure minimum risk of capacitive or inductive coupling.

Screening

The solution of the equation for an electromagnetic wave striking a conductor shows that the penetration of the wave into the conductor is attenuated according to the expression

$$I = I_0 \exp\left\{\frac{-x}{\delta}\right\} \qquad (6.103)$$

where $\delta = (\pi\sigma f\mu_0\mu_r)^{1/2}$. δ is a measure of the severity of the attenuation and is referred to as the *skin depth*. For a wave of frequency 1 MHz impinging on copper δ is approximately 6.6×10^{-3} cm. Clearly, a thin layer of metal foil is all that is required at these frequencies. Consequently, the use of conducting films or braiding, such as is found in co-axial cable, is a very efficient way to provide screening from high frequency electromagnetic interference. If a cage or mesh is used the size of the holes should be small compared with the wavelength.

Earthing

The obvious remedy to avoiding the problems associated with having multiple earths is to have a common earthing point to which all equipment is joined. This may be done as either a line or star arrangement as shown in figure 6.23.

Each technique has a disadvantage. The line approach could pose a health and safety risk if the earth connections between units become disconnected while the units were still connected to the mains. The star approach still has an antenna-type loop providing an opportunity for electromagnetic waves to enter the system.

It is now a mandatory safety requirement that mains powered electrical equipment has a connection to earth using an appropriate gauge of wire for the anticipated fault condition and that the installation, as a whole, is earth bonded. *Equipotential bonding* is the wiring together of the electrical mains earthing point, the water and gas supply pipes and any other ducting or piping systems at the points of entry to the facility and any exposed metallic parts of the building structure. Because of these constraints it may not be possible or desirable to adopt the strategies of figure 6.23. There are a number of techniques such that, even though the sensor and receiver are earthed, the signal lines are prevented from injecting any earth fault potential difference into the data.

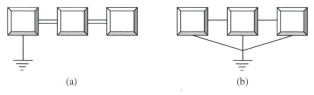

(a) (b)

Figure 6.23. Earthing strategies: (a) line, (b) star.

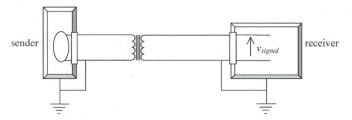

Figure 6.24. The isolation transformer.

(a) The isolation transformer

For optimum performance there should be a screen (figure 6.24) between the primary and secondary windings so as to reduce stray capacitance coupling. It is also possible to limit power transfer by running the core at near magnetic saturation, which may be an additional benefit.

This solution is effective but disadvantages are cost, physical size, d.c. isolation and poor frequency response.

(b) Balancing

The use of a centre-tap on the source signal and at the receiver circuitry effectively produces a differential system independent of earth potential. Because both conductors are part of a balanced system (figure 6.25) any common mode interference should be equal in both arms and so cancel at the receiver.

The drawbacks are that the source has to be in two parts, which may only be achievable using a centre-tap transformer with its inherent limitations, and that the component values and wiring impedances are identical in the two sides of the system. This latter condition becomes increasingly more difficult to achieve at higher frequencies due to stray capacitance.

(c) The differential amplifier

The differential amplifier (figure 6.26) has no input reference zero but uses its supply ground as zero. The amplifier power supply is not connected to the

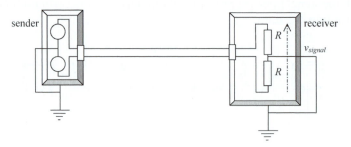

Figure 6.25. The balanced system.

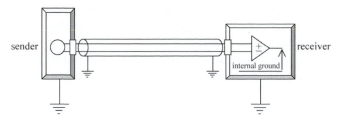

Figure 6.26. The differential amplifier system.

external earthing. Special twin-core co-axial cable is used. In applications where the sender unit has an internal ground that is required to be connected to the receiver's internal ground double screened, twin-core coaxial cable is used.

(d) Optical coupling

Optical isolation (figure 6.27) is the ultimate in providing separation from sender and receiver. Optical isolation is most easily accomplished with digital signals. For analogue applications care has to be taken over the electronic circuit design to ensure linearity.

Printed circuit board design

The problems highlighted in the previous section can be minimized by good design practice.

- Large tracks should be used for the power supply rails and any ground rail or plane to minimize resistance and therefore any drop in potential.
- Separate grounds should be used for analogue and digital sections for systems containing mixed components.
- Capacitors should be placed across the supply lines close to each integrated circuit (special miniature types are available that can be fitted below the chip) to steady their supply. A large capacitor is placed across the incoming

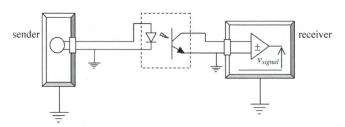

Figure 6.27. Optical isolation.

supply lines to ensure fast charging of the smaller capacitors. Tantalum electrolytic or metallized polycarbonate are the preferred type due to their low internal inductance.

- For high frequency applications miniature bead inductors of about $10 \, \mu H$ should be placed in the power supply lines to the high frequency sections of the system.
- Short leads are everywhere preferable so as to minimize electromagnetic reception and emission. Generally, all wire pairs should be twisted so as to nullify antennae behaviour.

Filtering

Some sensors need to be driven by an a.c. supply or have an a.c. signal conditioning circuit. For these systems it is possible to arrange for the a.c. frequency to be well away from any interference. Then a suitable filter can be used, similar to those described in the noise reduction section above, to allow only the desired signal through.

When interference is at a very specific frequency, such as the power line, a narrow band stop or *notch* filter is useful. A simple passive version is the twin-T filter (figure 6.28). The two sections of the T each cause a 90° phase shift, which are 180° out of phase at the notch frequency theoretically giving infinite attenuation. In practice attenuations are variable due to tolerances. See Horowitz and Hill (1989) for a comparison of filter types. Analogue filters are available as integrated circuits. These have better characteristics. The twin-T, for example, can provide a $-60 \, dB$ notch. The roll-off is only $-10 \, dB$ at $\frac{1}{2} f_{notch}$ and $2 f_{notch}$ and $-3 \, dB$ at $\frac{1}{4} f_{notch}$ and $4 f_{notch}$.

If the system contains a microprocessor then a digital filter could provide the optimum performance, removing tolerance and drift problems. These more advanced filters can be made to follow any drift of the interference frequency.

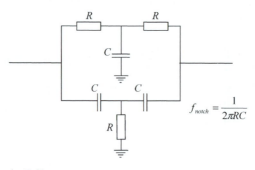

$$f_{notch} = \frac{1}{2\pi RC}$$

Figure 6.28. The twin-T filter.

References

Baher H 1991 *Analog and Digital Signal Processing* (Chichester: Wiley)

Beck M S and Plaskowski A 1987 *Cross-correlation Flowmeters* (Adam Hilger)

Bendat J S and Piersol A G 1986 *Random Data Analysis and Measurement, Procedures* 2nd edition (Wiley)

Bleaney B I and Bleaney B 1965 *Electricity and Magnetism* 2nd edition (Oxford University Press)

Brook D and Wynne R J 1988 *Signal Processing: Principles and Applications* (Edward Arnold)

Cooley J W and Tukey J W 1965 'An algorithm for the machine calculation of complex Fourier series' *Math. of Computation* **19** 297–330

Horowitz P and Hill W, 1989 *The Art of Electronics* 2nd edition (Cambridge University Press)

Kodali V P 1996 *Engineering Electromagnetic Compatibility* (New York Institute of Electrical and Electronics Engineers)

Meade M L 1983 *Lock-in Amplifiers; Principles and Applications* (London: Peter Peregrinus)

Nicollian E H and Brews J R 1982 *MOS Physics and Technology* (New York: Wiley)

Ott H W 1988 *Noise Reduction Techniques in Electronic Systems* 2nd edition (Wiley–Interscience)

Perls T A 1952 *J. Appl. Phys.* **23** 674

Wilmshurst T H 1990 *Signal Recovery from Noise in Electronic Instrumentation* 2nd edition (Adam Hilger)

Exercises

1. Distinguish between
   (i) deterministic and random signals
   (ii) noise and interference.

2. An ensemble of ten identical thermometers, each having a resolution of $0.1\,°C$, is placed in a large, thermally insulated tank of water which is continuously stirred. The tank is being steadily heated.
   (i) What class of system is this?
   (ii) If the tank were not being heated what would the classification be then?

3. A signal has a peak value of $2.0\,V$. Calculate its r.m.s. value if the signal is
   (i) a sinusoid,
   (ii) a saw-tooth,
   (iii) a square-wave of 40% duty cycle.
   In each case assume that the signal is unipolar and that the period is $20.0\,ms$.

4. The following diagram (figure E.2) shows the probability distribution $p(v)$ of the amplitude levels of a random signal $v(t)$.

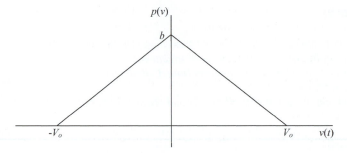

Figure E.2.

(i) Find the value of b in terms of V_0 so that $p(v)$ is correctly normalized.

(ii) Determine the mean and standard deviation of the signal.

(iii) Sketch the cumulative distribution function.

(iv) Estimate the probability that $v > \frac{1}{2}V_0$.

5. The auto-spectral density of a signal is shown below in figure E.5. Determine the autocorrelation function.

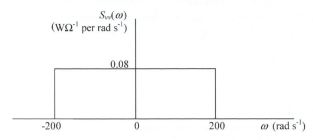

Figure E.5.

6. One of the rotor bearings in a helicopter is suspected of having wear. The normal output of a vibration sensor on the bearing housing is band-limited white noise. If wear has occurred this will contain an additional periodic signal of much smaller amplitude than the noise signal. Explain how confirmation of wear might be achieved and an estimate obtained of the signal amplitude.

7. A certain transducer is generating a sinusoidal signal of amplitude 0.1 V. The signal is corrupted with band-limited white noise occupying the band 0 to 200 rad s^{-1} and of amplitude 5 mW per rad s^{-1}. The frequency of the sinusoidal signal is at the centre of the noise bandwidth. Determine the auto-correlation of the signal and noise.

8. Why is twin core co-axial cable preferred to single core when connecting microphones in public address systems?

9. A certain thermocouple produces $39.4\,\mu\text{V}\,\text{K}^{-1}$. Given that the signal conditioning amplifier has an input impedance of $10.0\,\text{M}\Omega$ and a bandwidth of $100.0\,\text{Hz}$ estimate the minimum detectable temperature change if the ambient temperature is $300\,\text{K}$.

10. Explain how averaging can increase the signal-to-noise ratio. Describe both digital and analogue approaches to signal averaging. What is a lock-in amplifier?

11. The telemetry signal from a Hall effect probe is found to be corrupted because of the arcing of an a.c. electric motor nearby. Describe the spectrum of this electromagnetic pollution and suggest some methods of improving the relative signal strength.

12. A seismometer is placed by the track of a tram to monitor vibration as the vehicle passes. The tram has an over-head a.c. supply and the circuit return is through the metal rails. The seismometer consists of an open-framed mass-spring unit and variable reluctance pick-up. A single core co-axial cable carries the signal to a data logger placed on the ground 10 m away. When the tram passes as well as the anticipated seismic signal other random and deterministic signals are also present.
    (i) Explain the origins of these unwanted signals.
    (ii) Describe methods by which the relative signal strength might be improved.

Chapter 7

Novel developments in flow measurement systems

7.1 Introduction

In the preceding chapters the design of instrumentation systems, new types of sensors, signal recovery and processing have all been described. In this chapter a specific field of application is presented incorporating some of these topics, particularly that of the processing of the recovered signal. It is hoped that by examining a specific field of measurement the reader will appreciate the different approach taken when designing an overall system for a specific task rather than the analysis and development of a component element. This is the so-called *top-down* approach. Issues to be considered at the outset of a design may include

- the range of the variable to be measured,
- the precision of the final result needed,
- the system block structure as described in section 1.3,
- the numerical specification of each block in the structure,
- the overall transfer function as described in section 1.4,
- the contributory and overall error estimates and uncertainties,
- the electromagnetic compatibility and other safety issues.

Flow measurement

The most commonly measured variables are mass, temperature, length, flow rate, pressure and level in tanks and hoppers. Commercially, flow measurement is one of the most significant. There are more than 100 different techniques used to measure flow rate. In the UK there are over 200 suppliers of flow monitoring equipment. The supply and maintenance of flow measuring devices is a major industry. Excluding domestic flowmeters, annual sales are typically,

UK	30 000 units
Western Europe	150 000 units

USA	220 000 units
Rest of the world	180 000 units

There are in the region of 15 million installed working meters worldwide. Annual sales in the UK are worth £90 million, and worldwide sales are estimated at £1000 million.

Fluid dynamics

To appreciate the need for the variety of signal processing techniques used in modern flowmeters it is helpful first to understand the physical principles on which these flowmeters are based. The design of flowmeters involves an understanding of fluid dynamics, a discipline founded in classical dynamics and thermodynamics. The remainder of this section contains some terminology and definitions used commonly in fluid dynamics and the following section briefly describes the types of flow field that are encountered.

Fluid particles

The concept of a fluid *particle* is often used. This is not an atom or molecule of the fluid but rather some small fluid element that remains identifiably intact along the flow.

Viscosity

Unlike a solid, there are no shear forces in a fluid at rest. A fluid can be defined as that state of matter which cannot sustain a shear force when at rest. It has been observed experimentally that all fluids, however small their viscosity and even those that do not wet the boundary wall, have a stationary surface or *boundary layer* adjacent to the wall. If another layer of fluid slides against this boundary layer, friction occurs because of the viscosity of the fluid. This friction causes a shear force across some fluid element as shown in figure 7.1.

The boundary wall need not be involved. Shear forces occur whenever different layers of viscous fluid slide against each other. For flow in one dimension the shear force τ between adjacent layers is given by

$$\tau = \mu \frac{\mathrm{d}u}{\mathrm{d}y} \tag{7.1}$$

Figure 7.1. Shear force in one-dimensional flow.

where du/dy is the velocity gradient at the point being considered and μ is the absolute or dynamic viscosity. If μ is independent of du/dy the fluid is said to be *Newtonian*.

Pathline

This is the trajectory taken by a fluid particle.

Streamline

This is a line drawn at an instant in time through the fluid showing the local direction of the flow at all points up and down the stream. A fine chord fixed at an up-stream point would indicate one streamline. Many such streamlines can be constructed.

Stream tube

This is a closed surface formed at any instant in time by joining a set of adjacent streamlines. In a closed conduit the boundary wall is one of a set of stream tubes that can be drawn. Fluid cannot, by definition, move through the surface of a stream tube, making it a useful theoretical concept.

Steady flow

When the velocity and fluid properties at any given location are independent of time the flow is said to be *steady*. The streamlines no longer change position with time. Pathlines and streamlines become identical.

Discharge (Q)

This is the volume of fluid passing through a cross-section per second and is measured in $\mathrm{m^3\,s^{-1}}$.

Mean velocity ($\bar{u}$)

This is the discharge Q divided by the cross-sectional area A at any section and is measured in $\mathrm{m\,s^{-1}}$.

Uniform flow field

If the velocity is constant throughout the flow, neither time-dependent nor spatially-dependent, the flow is said to the uniform. It is possible to have steady non-uniform flow. An example of this is the steady flow of a fluid through a tapering tube.

The equation of continuity

The equation of continuity is derived from the concept of the *conservation of mass*. Once transients have died away the mass flow rate must be constant everywhere so in simple terms,

$$\rho Q = \text{constant} \tag{7.2}$$

or

$$\rho A \bar{u} = \text{constant}. \tag{7.3}$$

The general form of the equation of continuity for three-dimensional, time-dependent analysis can be expressed as

$$\frac{\partial \rho}{\partial t} + \frac{\partial(\rho u)}{\partial x} + \frac{\partial(\rho v)}{\partial y} + \frac{\partial(\rho w)}{\partial z} = 0 \tag{7.4}$$

where u, v and w are the velocity components in the x, y and z directions respectively. See Fox *et al* (2004). For the special case of an incompressible fluid $\rho = \text{constant}$, so equation (7.4) becomes

$$\frac{\partial u}{\partial x} + \frac{\partial v}{\partial y} + \frac{\partial w}{\partial z} = 0. \tag{7.5}$$

For steady flow there is no time-dependent variation of any fluid property. Equation (7.4) then becomes

$$\frac{\partial(\rho u)}{\partial x} + \frac{\partial(\rho v)}{\partial y} + \frac{\partial(\rho w)}{\partial z} = 0. \tag{7.6}$$

Bernoulli's theorem

Daniel Bernoulli published his work on fluid dynamics in 1738. Assuming that the fluid is ideal, meaning incompressible and having zero viscosity, and making the further assumption that the fluid flow is steady and irrotational Bernoulli's theorem can be stated as

> 'The sum of the pressure, plus the kinetic energy per unit volume, plus the potential energy per unit volume at a point on a streamline is constant'

The theorem can be confirmed by considering the mechanical energy changes experienced by a fluid particle flowing within some stream tube and assuming no energy loss. From figure 7.2, as the fluid flows through δx_a at position a work is done,

$$F_a \delta x_a = P_a A_a \delta x_a = P_a A_a u_a \delta t \tag{7.7}$$

where δt is the time taken to traverse δx. Similarly, flowing through δx_b at position b,

$$F_b \delta x_b = -P_b A_b u_b \delta t. \tag{7.8}$$

P_b is the pressure at the stream tube exit and so is acting to oppose the fluid motion hence the negative sign. The total work done is therefore

$$\Delta E = P_a A_a u_a \delta t - P_b A_b u_b \delta t. \tag{7.9}$$

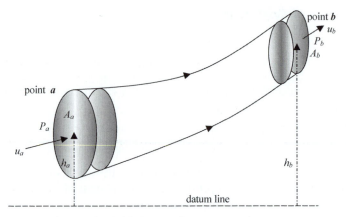

Figure 7.2. Energy changes for fluid elements in a stream tube.

This net work done is equivalent to the change in potential and kinetic energies of the fluid elements.

$$\text{Change in kinetic energy} = \tfrac{1}{2}m_b u_b^2 - \tfrac{1}{2}m_a u_a^2$$

$$= \tfrac{1}{2}\rho A_b \delta x_b u_b^2 - \tfrac{1}{2}\rho A_a \delta x_a u_a^2 \qquad (7.10)$$

$$\text{Change in potential energy} = m_b g h_b - m_a g h_a$$

$$= \rho A_b \delta x_b g h_b - \rho A_a \delta x_a g h_a. \qquad (7.11)$$

Adding equations (7.10) and (7.11) and equating to equation (7.9), with the provision that for an incompressible fluid $A_a u_a = A_b u_b$, gives

$$P_a + \rho g h_a + \tfrac{1}{2}\rho u_a^2 = P_b + \rho g h_b + \tfrac{1}{2}\rho u_b^2. \qquad (7.12)$$

This can be stated more generally that, for any point on a streamline,

$$P + \rho g h + \tfrac{1}{2}\rho u^2 = \text{constant} \qquad (7.13)$$

which is the Bernoulli equation. A more rigorous treatment involves Euler's equation; see Fox *et al* (2004).

P is sometimes referred to as the static pressure and the term $\tfrac{1}{2}\rho u^2$ as the dynamic pressure. Contrary to popular intuition, equation (7.13) states that for horizontal flow if the pressure decreases the velocity increases.

This simple relationship can be used to explain many common phenomena such as the pressure drop inside a railway carriage on entering a tunnel, aerodynamic lift, the trajectory of a spinning ball and the bow wave produced in front of a moving ship. It is also the basis of the operating principle of constriction-type flowmeters such as the orifice plate and Venturi tube and of the Pitot-static tube; see Bentley (1995).

7.2 Types of flow field

Sometimes the smoke from a cigarette appears to show two distinct types of flow. Near the source the smoke column appears smooth and regular. Farther away sections within the flow become separately identifiable, the column becoming irregular, twisting and turning.

It was Osborne Reynolds who investigated these two types of flow in some detail. He did this by injecting dye into water flowing in a transparent pipe. At relatively low flow velocities the dye streak follows a line parallel to the pipe wall. This is termed *laminar* flow. At higher flow velocities the particles have random, fluctuating velocities. This is termed *turbulent* flow. In 1883, Reynolds noted that the change between flow types did not just depend on the flow velocity, but on the empirically derived expression,

$$Re = \frac{ud}{\eta} \tag{7.14}$$

where d is the pipe diameter and η the kinematic viscosity (defined as $\eta = \mu/\rho$) and Re the Reynolds number. Although the Reynolds number was devised for flow régimes in pipes it can be adapted for other flow applications. It is in effect a measure of whether inertial forces or viscous forces dominate within a particular flow field.

7.2.1 Laminar flow

Laminar flow is said to occur when the fluid particles move in layers or *laminae*. Fluid particles close to a large plate will move in a plane parallel to the plate surface and parallel to each other. Their velocity will depend on their distance from the plate. In a circular pipe such laminae form a series of co-axial cylinders extending from the pipe axis to the wall. This is sometimes referred to as Poiseuille flow. To obtain laminar flow the Reynolds number must be less than about 2300. When viscous forces dominate the flow is usually laminar.

The velocity profile in a circular pipe can be determined from considering the shear forces. Assuming equilibrium exists then the pressure gradient will have reached a steady value. The cylindrical section shown in figure 7.3 illustrates the relationship between pressure gradient and shear force.

At equilibrium the force provided by the pressure gradient across the faces of the cylinder is equal to the force from the sum of the shear forces on the cylinder sides,

$$\Delta P \, \pi r^2 = 2\pi r \, \Delta x \, \tau. \tag{7.15}$$

Equation (7.15) can be combined with equation (7.1) to give

$$\frac{du}{dr} = \frac{r}{2\mu} \frac{\Delta P}{\Delta x}. \tag{7.16}$$

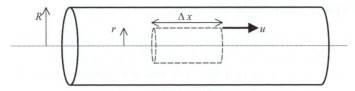

Figure 7.3. Shear force and pressure gradient.

Assuming a short cylinder so that $\Delta x \to \mathrm{d}x$, on integration over the pipe radius R with the boundary condition that $u = 0$ at the wall, equation (7.16) yields

$$u = -\frac{R^2}{4\mu}\left(\frac{\mathrm{d}P}{\mathrm{d}x}\right)\left[1 - \left(\frac{r}{R}\right)^2\right]. \tag{7.17}$$

Since the flow profile is axially symmetric and the maximum velocity occurs on the axis this can be re-written as

$$u = U_{\mathrm{max}}\left[1 - \left(\frac{r}{R}\right)^2\right]. \tag{7.18}$$

For an axially symmetric flow régime the mean velocity can be generally determined using

$$Q = A\bar{u} = \int_0^R u(r)2\pi r\,\mathrm{d}r. \tag{7.19}$$

For laminar flow this yields the useful relationship $\bar{u} = \frac{1}{2}U_{\mathrm{max}}$.

7.2.2   Turbulent flow

If the Reynolds number for the pipe is in excess of about 2300 then turbulent flow occurs. When inertial forces dominate the viscous forces the flow is usually turbulent. Fluid particles move randomly. Streamlines constantly change their configuration. Equation (7.1) does not apply when the flow is turbulent. Theoretical analysis of turbulent flows is difficult and formulae are often semi-empirical. For many turbulent flow régimes within pipes the time average velocity $\hat{u}$ at some point r from the tube axis is unchanging. Fox *et al* (2004) state that $\hat{u}$ can then be represented by

$$\hat{u}(r) = U_{\mathrm{max}}\left[1 - \frac{r}{R}\right]^{1/m} \tag{7.20}$$

where m depends on the Reynolds number. This generally models the empirical behaviour except at the axis and close to the wall. Assuming the profile is valid from $r = 0$ to R then equation (7.20) gives

$$\bar{u} = U_{max} \left[\frac{2m^2}{(m+1)(2m+1)} \right]. \tag{7.21}$$

For fully developed turbulent flow often $m \cong 7$ is taken as a representative value. Equation (7.21) then gives $\bar{u} = 0.82 U_{max}$. Expressing equation (7.20) in terms of $\bar{u}$ gives

$$\hat{u}(r) = \bar{u} \left(\frac{(m+1)(2m+1)}{2m^2} \right) \left[1 - \frac{r}{R} \right]^{1/m}. \tag{7.22}$$

Generally, if the Reynolds number is close to 2300 the type of flow can fluctuate between laminar and turbulent. However, experiments have been carried out where, using very smooth surfaces and isolating all disturbances, laminar flow has been maintained even when Re has reached 100 000. But for common use the value of 2300 is a reliable indicator for the expected flow régime in a pipe.

7.2.3 Non-steady flow

The theoretical modelling of non-steady flow is complex. Womersley (1957) developed a theory for oscillatory flow and pulsed transmission. The study was for the analysis of blood flow in mammalian arteries and was not the first attempt at a theoretical description for this type of flow.

Womersley's paper begins with an analysis of a viscous liquid flowing in a straight rigid circular tube. He assumed no radial motion so that u is not a function of the position along the tube axis x but only of the radial distance r from the axis. This infers that the pressure gradient $\partial P / \partial x$ is also independent of x. The acceleration is given as

$$\frac{\partial u}{\partial t} = -\frac{1}{\rho} \frac{\partial P}{\partial x} + \eta \left(\frac{\partial^2 u}{\partial r^2} + \frac{1}{r} \frac{\partial u}{\partial r} \right). \tag{7.23}$$

In the first section of his paper Womersley used an oscillatory pressure gradient of the form

$$\frac{\partial P}{\partial x} = A \cos(\omega t - \phi). \tag{7.24}$$

The solution to equation (7.23) then involves Bessel functions. Using modulus and phase forms M_0' and ε_0 he obtained an expression for u of the form

$$u = \frac{AR^2}{\mu} \frac{M_0'}{\alpha^2} \sin(\omega t - \phi + \varepsilon_0) \tag{7.25}$$

where R is the tube radius and α is a non-dimensional parameter. The value of α is said to be useful in characterizing the flow. It is given by

$$\alpha = R\sqrt{\frac{\omega}{\eta}}. \tag{7.26}$$

The flow velocity is reduced by the viscosity but also depends on the value of α.

In the following sections of his paper Womersley analysed flow in a freely moving elastic tube and then went on to develop modifications for additional mass, elastic constraint and internal viscous damping. In the final section of the paper he discusses the issue of having a thin boundary layer of lower viscosity. This was considered to be a significant issue for blood flowing in small vessels. For a fuller description of the mechanics of blood flow see Evans *et al* (1989).

Flow régimes and flowmeters

Laminar flows have a lower resistance to flow and so are preferred in some situations such as long pipelines in order to give greater pumping efficiency. In heat exchangers and with oxygen transfer in blood flow, turbulence gives good mixing and so improves the transfer process. Some flowmeters will not function correctly in turbulent flows so precautions must be taken with these. Others only work well in turbulent flows because turbulence reduces the effects of viscosity. Also, for turbulent flow the variation in m only has a small effect on the estimation of $\bar{u}$ from $U_{\max}$, whereas upsets to a laminar flow can give errors of 30%. Flowmeters which can work in either type of flow are usually calibrated for one or the other which means that if the flow régime changes erroneous results occur.

Many flowmeters will not function correctly in swirling or pulsating flows. For such flowmeters manufacturers specify installation requirements such as having long straight runs of pipeline before and after the meter and inserting vane structures into the flow.

7.3   The cross-correlation flowmeter

The correlation technique has been used in applications where the more traditional flowmeters cannot operate well. The concept of correlation was described in section 6.3.4. In this context it is used to determine the flow velocity. Cross-correlation flowmeters have been developed for applications where conventional flowmeters will not function or would suffer damage. Velocity measurement has been achieved during the conveying of materials such as molten chocolate, slurries, powders, textiles and the flow of sewage.

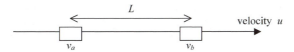

Figure 7.4. Sensors spaced along some velocity field.

Two sensors are used, spaced at intervals along the velocity direction as illustrated in figure 7.4. Any type of sensor can be used as appropriate for the flow application.

The essential requirement is that the process has some randomly fluctuating parameter such as temperature, conductivity, opacity, electrostatic charge, dielectric constant or density. Ideally, the local value of this parameter should not change significantly between the sensors so that the same random signals v_a and v_b are produced at both sensors but simply time-shifted. This *frozen pattern* is illustrated in figure 7.5.

Using the cross-correlation procedure on the signals v_a and v_b a peak in the cross-correlation function $R_{v_a v_b}$ occurs where the variable τ is equal to the time of travel T between sensors (figure 7.6).

The velocity is determined using L/T. Obviously, large values of T and L would tend to give a more precise estimate of the flow velocity. However, the height of the correlation peak is reduced as the correlation between the two sensor signals deteriorates. This sets a limit on the sensor spacing L.

Early cross-correlation systems were analogue and consequently prone to zero drift. The preferred approach is to use digital technology. The correlation algorithm is derived from that given in section 6.3.4 repeated below,

$$R_{v_a v_b}(q\Delta t) = \frac{1}{N}\sum_{j=1}^{N} v_{aj} v_{b(j+q)} \tag{6.44}$$

where Δt is the clock period. The time shift τ is $q\,\Delta t$. The number of sampling points is N for each of the time windows for v_a and v_b. Consequently, there are N calculated values for the function $R_{v_a v_b}$. This number, in conjunction with the clock speed, has to be sufficient for the correlation to occur, i.e.

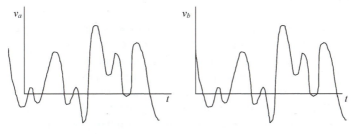

Figure 7.5. Sensor output signals.

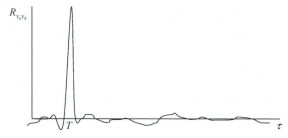

Figure 7.6. The cross-correlelogram of v_a and v_b.

$N \Delta t > T$. This implies that a larger clock period is preferable for lower velocities otherwise large allocations of memory are needed and the algorithm will take longer to run. This conflicts with system performance. Beck and Plaskowski (1987) give a relationship for the fractional velocity resolution as

$$\frac{\Delta u}{u} = \pm \frac{\Delta t}{2L} u. \qquad (7.27)$$

Larger clock periods mean poorer resolution. Also, resolution is worse at higher velocities. In designing the digital system the following issues need to be considered:

- What is the maximum value of L that can still give a correlation peak above noise in the correlelogram?
- What is the dynamic range of velocity likely to be encountered?
- What resolution is desired for the maximum velocity?

It is interesting to note that the wider the bandwidth of the random parameter that is being recorded in the signals v_a and v_b the lower the level of the background noise in the function $R_{v_a v_b}$. This means that the more random the flow parameter providing the signals v_a and v_b the better the performance of the system.

7.4   Ultrasound

Although ultrasound is defined as sound beyond the human range of about 20 kHz, for flow measurement frequencies in the region of 1 MHz are commonly used. This is in order to achieve spatial resolution of about 1 mm. For any wave-based system resolution cannot be greater than the wavelength of the wave employed. Because the speed of mechanical vibrations in solids and aqueous liquids is of the order 3 km s^{-1} and 1.5 km s^{-1} respectively the required small wavelengths mean correspondingly high frequencies. This is inconvenient in that air is opaque to ultrasound at these frequencies. Consequently, a coupling gel or compound has to be used between the transducers and the target. Also, any large bubbles or gas cavities in the beam path will obscure the measurement. (This is a useful feature long-used for the detection of flaws in welds and castings.)

Ultrasonic transducers

A number of materials have been used for transducers. Due to the high frequencies involved materials having piezoelectric properties are found to be the most appropriate. These include quartz, ceramics such as lead zirconate-titanate, lead metaniobate and barium titanate and the plastic film polyvinylidene fluoride (PVDF). Quartz is less commonly used now. PVDF has a closer mechanical impedance match to aqueous media than do the other materials enabling better coupling; see section 1.6. (The paper by O'Donnell *et al* (1981) includes data on various piezoelectric materials.) Because of this and its flexibility, PVDF could be used for direct contact with the human body without the need for coupling gel. Unfortunately, it is a relatively weak transmitter compared with ceramics operated at resonance. Consequently, ceramics are the most commonly used for these high-frequency transducers. The coupling medium used to exclude air between the transducers and target can also act as an impedance matching device.

For ultrasonic applications transducers are used in either pulsed mode or continuous working (CW) mode. The equivalent circuit of a piezoelectric transducer is shown in figure 7.7. In section 2.2.2 the concept of the driven oscillator was described and in section 2.2.3 the Q factor. As mentioned there, high-Q devices have the benefits of narrow bandwidth, low loss per cycle and greater immunity to loading effects. For CW systems, having a stable narrow bandwidth is useful when performing frequency analysis on echo signals as described later. However, in pulsed systems high Q causes 'ringing' of the transducer after the end of the electrical pulse because of the low loss. This reduces the time resolution. It is possible to alter the mechanical Q of the transducer by loading the output signal with electrical

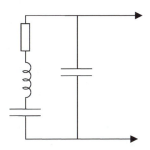

Figure 7.7. The equivalent circuit of a piezoelectric transducer.

impedance. In section 1.5 the analysis of the dynamic response of a system was described. In this way the piezoelectric element of figure 7.7 with an additional electrical impedance network could be modelled and the resultant pulse response simulated in a software package such as Matlab®. A minor disadvantage to reducing the Q would be some loss of signal amplitude.

Time-of-flight flowmeters

The measurement of flow using ultrasound has become an established technique in the process industries (figure 7.8). The time-of-flight approach uses separate transmit and receive transducers within the flow mounted on either side of the pipe. Acoustic coupling with the liquid is not a problem and the use of preformed housings means that the angle of incidence is precisely known.

Ultrasound is transmitted alternately P to Q and then Q to P. The velocity u varies along the length l according to some profile. Consequently, to correctly determine ΔT the mean value of velocity $\bar{u}$ must be used. Assuming a pencil ray of ultrasound, $\bar{u}$ across the pipe is determined using

$$\bar{u} = \frac{1}{R} \int_0^R u(r)\, dr. \tag{7.28}$$

The function $u(r)$ chosen will depend on the flow type. Equation (7.18) would be used for laminar flow or equation (7.21) for turbulent flow. The propagation times are

$$T_{PQ} = \frac{l}{c + \bar{u}\cos\theta} \tag{7.29}$$

and

$$T_{QP} = \frac{l}{c - \bar{u}\cos\theta}. \tag{7.30}$$

Combining gives

$$T_{QP} - T_{PQ} = \frac{l}{c - \bar{u}\cos\theta} - \frac{l}{c + \bar{u}\cos\theta} \tag{7.31}$$

$$\Delta T = \frac{2R}{\sin\theta}\left[\frac{1}{c - \bar{u}\cos\theta} - \frac{1}{c + \bar{u}\cos\theta}\right] \tag{7.32}$$

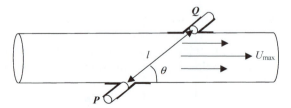

Figure 7.8. The time-of-flight flowmeter.

where $2R$ is the pipe diameter. Usually the speed of the ultrasound $c \gg \bar{u}$, so

$$\Delta T \cong \frac{4R\bar{u}\cot\theta}{c^2}. \tag{7.33}$$

For measurement of $\bar{u}$ to a resolution of 1% in a pipe of diameter 0.1 m with a flow velocity of $1\,\mathrm{m\,s}^{-1}$ requires the measurement of ΔT to a precision of $<1\,\mathrm{ns}$. The development of circuitry to deliver this is not a trivial task whether done as a true timing system, or as a phase-shift measurement of the received signal compared with that transmitted, or as a correlation system using a pseudo-random binary sequence. To avoid this difficulty the first ultrasonic flowmeters marketed were of the so-called *sing-around* type; see Sanderson and Hemp (1981). This is based on frequency rather than timing circuitry. Here the upstream and downstream signal times are used to provide the time periods for two separate oscillators. The difference frequency of the two oscillators is

$$\Delta f = \frac{\bar{u}\sin 2\theta}{2R}. \tag{7.34}$$

A further advantage of this technique is that the value of c is not required. Consequently, any drift due to temperature or change in liquid composition is not an issue. However, bubbles in the flow can cause loss of signal forcing the signal processing circuitry to re-start. Scott (1984) compares the performance of the two types under calibration.

7.5   The Doppler effect

Relative motion between a transmitter and receiver gives the well-known Doppler frequency-shift effect. It is possible to measure flow velocity non-invasively by transmitting the ultrasound into the flow field and then examining the frequency shift of the echo. For the system to work well there must be scattering particles within the fluid. The frequency is Doppler shifted twice. Firstly, the scatterers receive a Doppler-shifted frequency f_1 as moving 'observers' of a stationary transmitter of frequency f_0,

$$f_1 = f_0\left(1 + \frac{v}{c}\right) \tag{7.35}$$

where v is the velocity of the scatterer and c the velocity of the sound in the medium. This is then Doppler-shifted to f_2 since the scatterers now behave as moving transmitters emitting back to the stationary receiver outside the pipe. The Doppler shift for a moving transmitter and stationary receiver is given by

$$f_2 = \frac{f_1 c}{c - v} \tag{7.36}$$

which is slightly different from equation (7.35). This is because in the first case the transmitter and medium are both stationary. In the second situation the transmitters are moving through the medium. Combining both Doppler shifts gives

$$f_2 = \left(\frac{c+v}{c-v}\right)f_0. \tag{7.37}$$

If the ultrasonic beam is inclined at some angle θ to the flow then v is re-written as $v\cos\theta$. Also, since usually $c \gg v$ the difference frequency can be approximated to

$$\Delta f = 2f_0 \frac{v\cos\theta}{c}. \tag{7.38}$$

For example, a transmitter operating at 10 MHz, mounted at 30° to the flow axis, in an aqueous flow with particulates moving at a velocity of $0.9 \times 10^{-2}\,\mathrm{m\,s^{-1}}$ would produce a frequency shift of about 100 Hz.

The signal processing of Doppler echo systems is often achieved using techniques developed by telecommunication engineers. The Doppler echo signal corresponds to that of a single-sideband transmission system of carrier frequency 10 MHz modulated at 100 Hz. A simple way of recovering the 100 Hz from the 10 MHz is to multiply the echo by the original transmission frequency. This can be readily achieved using analogue multiplication. From the trigonometrical identity,

$$\cos A \cos B = \tfrac{1}{2}\cos(A - B) + \tfrac{1}{2}\cos(A + B) \tag{7.39}$$

$$\cos 2\pi f_1 t \cos 2\pi f_2 t = \tfrac{1}{2}\cos 2\pi \Delta f + \tfrac{1}{2}\cos 2\pi (f_1 + f_2)t. \tag{7.40}$$

The high-frequency term $f_1 + f_2 \cong 20$ MHz can be easily removed by filtering. Analogue multipliers are readily available in integrated circuit form.

Some early industrial systems using a non-invasive, Doppler approach did not live up to the claims promised by the instrument manufacturers. Users encountered two problems. Firstly, the attachment of the transducers to the outside of the pipe using gel or compound meant that sometimes the angle of incidence of the ultrasound beam with the pipe axis was not that determined during calibration. This was due to either poor installation or because of additional refraction caused by scaling on the internal pipe walls. Also, transverse waves can be established in solids whose velocities are different to those of longitudinal waves. Using the wrong velocity affects the calculated angle of refraction of the wave entering the flow. Secondly, if the flow régime changed, as described at the end of section 7.2, estimation of the mean velocity from the maximum Doppler-shifted frequency was incorrect. Gessner (1969) estimated the errors that might be expected when profiles changed. Cousins (1978) published calibration results for various sizes of pipe and flow régimes. Interrogation of the profile by multiple ultrasonic beams has been proposed as a means of obtaining a more reliable estimate of the mean velocity. Another approach is the use of a pulsed or 'gated' beam. In this way the mean velocity for a small zone of the fluid can be determined and hence the mean velocity along the line of the beam for the whole. However, pulsed systems require more power and are limited in spatial range and maximum flow velocity, see Sanderson and Hemp (1981).

7.5.1 The echo spectrum

Another approach to the problem of uncertain velocity profiles is to perform a frequency analysis on the echo signal. In this way an estimate of the range of velocities and, from the intensities of the spectral lines, the velocity distribution can be made. This procedure has become very popular in clinical flow measurement. Frequency spectra of the type shown below in figure 7.9 are easily produced using the FFT on the low frequency (demodulated) echo signals Δf. The symbol f_D is used to denote Δf in the text from here.

WORKED EXAMPLE

A certain signal has a bandwidth of 0–10 kHz. A frequency analysis is required and a FFT is to be used. If the frequency resolution is to be about 80 Hz determine the required sampling rate of the input signal, the length of the time window and the size of the FFT needed.

According to the sampling theorem described in section 5.2, the sampling rate must be at least twice the highest frequency present to avoid aliasing. (A system designed to accept 10 kHz would have an analogue input filter to prevent higher frequencies from entering. There would also be a filter for enhancement to reduce the effects of 'windowing'.) Therefore, **the sampling rate is $2 \times 10^4 \, \mathrm{s}^{-1}$**.

The length of the sampling window determines the frequency resolution. In section 5.2 it was shown that the effects of sampling produced a sinc function in the frequency domain. A similar effect occurs with the whole sampling window. The Fourier transform of the time window produces a sinc function that determines the sharpness of the spectral lines. The width of the frequency domain sinc function is approximately $1/T_W$ where T_W is the window time. For the spectral lines to be separated so as to give a frequency resolution of 80 Hz the time window must be about 12.5 ms or larger. The sampling rate is $2 \times 10^4 \, \mathrm{s}^{-1}$ so during this time 250 samples will be recorded for use by the FFT. However, the FFT algorithm requires that the number of input points be a highly composite number of 2^N; see Rabiner and Gold (1975). So, in reality 256 (2^8) samples must be taken. This **256-point FFT** will generate 128 unique frequency points in the output (a further 128 conjugate points are also produced that can be discarded).

The FFT sampling time will thus be $256/2 \times 10^4$ s giving a value for the **time window of 12.8 ms**. From above, the constraint of choosing 2^N means that the actual frequency resolution will be about 78 Hz.

The mean frequency of figure 7.9 can be simply determined using

$$\bar{f}_D = \frac{\int_0^\infty \rho(f_D) f_D \, df_D}{\int_0^\infty \rho(f_D) \, df_D} = \frac{\sum_{i=0}^{N} w_i f_{Di}}{\sum_{i=0}^{N} w_i} \tag{7.41}$$

where $\rho(f_D)$ is the intensity distribution and w_i are the weightings of the discrete form of a spectrum of N frequency points. If the assumption is made that the intensity of a received frequency component is simply proportional to the number of scatterers travelling with the same velocity the mean velocity can be estimated. Using equation (7.38),

$$\bar{f}_D = \frac{\sum_{i=0}^{N} w_i f_{Di}}{\sum_{i=0}^{N} w_i} = \frac{2 f_0 \cos \theta}{c} \bar{u}. \tag{7.42}$$

Brody (1974) states that, under these idealized conditions, knowing the pipe cross-sectional area A equation (7.42) should give the volume flow rate Q using $Q = A\bar{u}$ and that this does not assume *a priori* knowledge of the flow profile.

However, estimation of the mean velocity is not quite as straightforward as it might first appear. Figure 7.10 for a laminar flow régime illustrates the difficulty of the problem. In a more rigorous analysis the following issues have to be considered:

- the effect of having a non-parallel beam,
- uncertainty in the angle of insonation,
- the effects of multiple scattering,

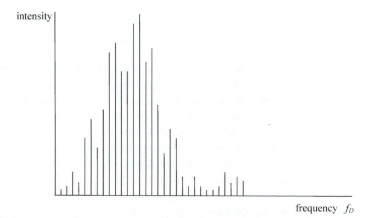

Figure 7.9. A typical frequency spectrum from a Doppler flowmeter.

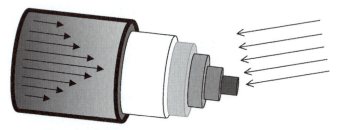

Figure 7.10. Insonation of a velocity profile.

- the distribution of the scatterers,
- the attenuation of the beam through the flow along the transmit and echo paths.

Angelsen (1980) has developed a theoretical analysis. Evans *et al* (1989) also describe work done in the field.

However, notwithstanding the theoretical difficulties of developing a model for the scattering process, it has been found that the frequency spectra can be empirically related to the mean velocity and good repeatability has been reported.

7.5.2 Forward and reverse flow

A further very useful feature that can be obtained from the Doppler-shifted echo spectrum is the identification of fluid moving toward or away from the transducer system. In the process industries this is not a common requirement. However, in the medical field it can be extremely useful. For example, reverse blood flow can occur around leaky heart valves.

The echo signal can be represented as

$$v_e = \sum_i A_i \cos(\omega_0 t + \phi_i) + \sum_j F_j \cos(\omega_0 + \omega_{F_j})t + \sum_k B_k \cos(\omega_0 - \omega_{B_k})t$$

$$(7.43)$$

where A_i is the amplitude of an echo of the original transmitted frequency ω_0, F_j is the amplitude of the echo received from scatterers moving towards the receiver and B_k the amplitude of the echo from a scatterer moving backwards. In practice the echo will be a continuous spectrum but performing the FFT as described above will generate discrete spectral lines.

As already described, obtaining the frequency-shifted components is relatively straightforward if only the frequency shift is required. In order to determine whether the frequency is up- or down-shifted more complex signal processing is necessary. Nippa *et al* (1975) list some options. These are considered below. If the flow velocity is in the range $0.9 \times 10^{-2}\,\mathrm{m\,s^{-1}}$ to $0.9\,\mathrm{m\,s^{-1}}$ then for a 10 MHz system the frequency shift will be between

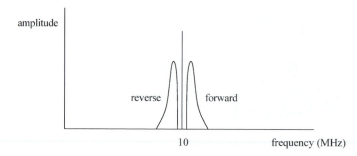

Figure 7.11. Echo spectrum for forward and reverse flow.

100 Hz and 10 kHz. The spectrum is illustrated in figure 7.11, although not to scale.

(1) Separation by filtering directly

It might be thought that simple filtering of the incoming echo spectrum would be a suitable approach. The received components will be 10 MHz, 10.0001–10.001 MHz and 9.9999–9.99 MHz. However, as Nippa *et al* (1975) point out the separation between frequencies in the region of 10 MHz and 10.0001 MHz to the required 40 dB is an impossible task using filters especially when considering that the transmit frequency will drift.

(2) Frequency shifting

Down-shifting of the Doppler spectrum means that the performance required of the filters becomes less demanding. Frequency shifting is a common telecommunications procedure. For example, the composite stereo signal in the UK, VHF-FM radio broadcasting uses frequency shifting to make better use of the transmission bandwidth.

Frequency shifting is again achieved by a multiplication procedure. The procedure used here is referred to as *heterodyning* by radio engineers. A frequency ω_m, derived from the transmission frequency but slightly lower, is multiplied with the echo signal and sum and difference terms are generated in the usual fashion. The frequency used for multiplication, ω_m, must be such that after the multiplication the difference term places the echo bandwidth in a suitable range in the lower part of the frequency spectrum (figure 7.12).

A phase-locked loop technique can be used to generate ω_m; see Horowitz and Hill (1989). The value of ω_m can be expressed as

$$\omega_m = \omega_0 - \omega_{het} \tag{7.44}$$

where ω_{het} is generated by a fixed, low-frequency oscillator. Because ω_m is derived form ω_0 any drift in ω_0 will not cause the recovered signals to drift. Clearly, ω_{het} must be greater than the highest Doppler frequency expected.

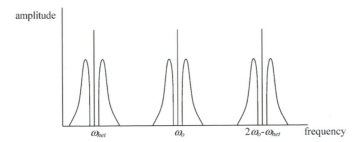

Figure 7.12. Echo spectrum for forward and reverse flow.

After filtering out the high frequencies there will be two spectral bands,

$$\omega_{\text{het}} - \sum_j \omega_{F_j} \quad \text{and} \quad \omega_{\text{het}} + \sum_k \omega_{B_k}$$

and the spectral line ω_{het}.

A further very sharp, notch filter could be used to remove ω_{het} but a modern implementation of this technique would be to use dsp at this stage rather than analogue techniques. Performing a FFT on the now low-frequency signal would give the spectrum directly and ω_{het} ignored.

(3) Phase rotation

Because of the demands placed on analogue filters by the two previous methods the major part of the paper by Nippa *et al* (1975) reports work on a phase-shifting system. The technique is similar to that of phase-quadrature detection (figure 7.13) used in telecommunications engineering. It involves two, precise, 90° phase-shifting elements as shown below.

For convenience a single velocity component from the echo spectrum of equation (7.43) is used for illustration,

$$v_e = A\cos(\omega_0 t + \phi) + F\cos(\omega_0 + \omega_F)t + B\cos(\omega_0 - \omega_B)t. \tag{7.45}$$

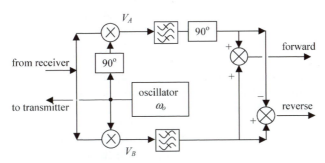

Figure 7.13. Phase-quadrature detection.

Multiplying the echo signal by the phase-shifted transmission frequency $D\cos(\omega_0 t + \pi/2)$ gives

$$V_A = D\cos\left(\omega_0 t + \frac{\pi}{2}\right)A\cos(\omega_0 t + \phi) + D\cos\left(\omega_0 t + \frac{\pi}{2}\right)F\cos(\omega_0 + \omega_F)t$$

$$+ D\cos\left(\omega_0 t + \frac{\pi}{2}\right)B\cos(\omega_0 - \omega_B)t. \tag{7.46}$$

Using the trigonometrical identity of equation (7.39) and filtering to remove high frequency and d.c. components gives,

$$V_A = \frac{1}{2}DF\cos\left(\frac{\pi}{2} - \omega_f t\right) + \frac{1}{2}DB\cos\left(\frac{\pi}{2} + \omega_B t\right) \tag{7.47}$$

or

$$V_A = \tfrac{1}{2}DF\sin\omega_F t - \tfrac{1}{2}DB\sin\omega_B t. \tag{7.48}$$

But the signal V_A is further phase-shifted by 90° so equation (7.48) becomes

$$V'_A = \frac{1}{2}DF\sin\left(\frac{\pi}{2} + \omega_F t\right) - \frac{1}{2}DB\sin\left(\frac{\pi}{2} + \omega_B t\right) \tag{7.49}$$

which on simplifying gives

$$V'_A = \tfrac{1}{2}DF\cos\omega_F t - \tfrac{1}{2}DB\cos\omega_B t. \tag{7.50}$$

Correspondingly, multiplying the echo signal by the transmission frequency $D\cos\omega_0 t$ gives

$$V_B = D\cos\omega_0 t A\cos(\omega_0 t + \phi) + D\cos\omega_0 t F\cos(\omega_0 + \omega_F)t$$

$$+ D\cos\omega_0 t B\cos(\omega_0 - \omega_B)t. \tag{7.51}$$

Simplifying and filtering as before, this reduces to

$$V_B = \tfrac{1}{2}DF\cos\omega_F t + \tfrac{1}{2}DB\cos\omega_B t. \tag{7.52}$$

The outputs are then

$$V'_A + V_B = DF\cos\omega_F t \tag{7.53}$$

$$V_B - V'_A = DB\cos\omega_B t. \tag{7.54}$$

Two important constraints for the system to work well are:

- The amplitudes DB in the signals V'_A and V_B must be of the same magnitude for the addition and subtraction procedures of equations (7.53) and (7.54) to work correctly. Similarly for the amplitudes DF. This will require some form of adjustable gain amplifier within the system. The signals in the system by Nippa *et al* (1975) differed by less than 0.2 dB.
- The two, 90° phase-shifters must perform well over the range of frequencies. The high frequency phase-shifter does not have a relatively large frequency

spread so is the less demanding to design. The low frequency second phase-shifter covers a large range. According to Nippa *et al* (1975) the design used in their system was an eight-pole, transistor-based filter which achieved $90° ± 0.6°$ over their range of 50 Hz to 7.5 kHz. A circuit published by Dickey (1975) uses op-amps to generate a 90° phase-shift for the range 100 Hz to 10 kHz.

Because of the advances of dsp, in a modern implementation the low frequency sections of the system, filtering, gain-adjust, phase-shifting, addition and subtraction would be processed digitally. Digital systems are more prescriptive to design. They are also very stable in operation because performance is not dependent on component values unlike analogue systems which drift with age and temperature.

7.5.3 Blood flow measurement

The measurement of blood flow-rate is of significant importance in a number of clinical specialisms. However, making accurate, non-invasive measurements of true volume flow-rate is not straightforward. Some clinical areas where this would be useful are given below.

- In order to assess the performance of the heart the blood volume flow-rate is required. Currently, the dilution method is routinely used; see Khandpur (1987). Here iced water is injected into the artery and the resultant temperature change used to determine the dilution and hence the volume. Clearly, as with any invasive procedure, it is distressing to the patient and not without risk.
- In the development of the unborn baby, as the child grows the umbilical supply has to increase to provide a greater supply of oxygen and nutrients. If the chord is impaired the child causes the blood pressure of its mother to increase in an attempt to obtain the desired supply rate. This high blood pressure is a sign of the condition know as pre-eclampsia and can be dangerous to the mother and baby. The use of a simple ultrasonic probe can detect velocities but not total volume flow rate.

Some areas of blood flow measurement do not require total volume flow rate but only some indication of change in flow velocity profile.

- The partial blockage caused by a thrombosis results in much higher flow velocities near the obstruction. A simple, hand-held ultrasonic transducer with audio frequency output can be used to locate the site of the blood clot.
- The growth of a tumour is marked by a stage where, in order to sustain growth, a vascular system within the tumour has to develop. Wells *et al* (1977) reported work on Doppler-shifted signals arising from microcirculation within malignant breast tumours. The structure of new vessels in tumours is different from normal tissue in that diameters are much

larger, walls are thinner and lack contractile elements. Burns *et al* (1982) reported that Doppler-shifted spectra from blood flowing in and near a breast tumour are characteriztically different and that this might be a useful diagnostic procedure.

Ultrasonic imaging systems are now very well developed. Duplex systems not only produce an image but can also perform Doppler-shift measurements on the imaging zone at a location set by a cursor superimposed on to the image displayed on the monitor. Some duplex systems colour code the image so that flow detected by the Doppler-shift appears as shades of red or blue in an otherwise monochrome image. In addition green can be added as a function of the variance of the signal as an indication of turbulence; see Namekawa *et al* (1982). In this way clinicians can see whether flow is towards or away from the probe and if turbulence is present green mixed with red or blue gives shades of yellow or cyan respectively.

It might be thought that, with a sophisticated duplex imaging system, a reliable estimate of the volume flow-rate should be possible by measurement of blood vessel diameter and measurement of mean flow velocity using Doppler-shift. Unfortunately, in addition to the problems of obtaining a reliable mean velocity estimate from the echo as described above, a number of other difficulties exist:

- The tube may not be circular.
- The tube diameter may vary during systole and diastole.
- The type of flow régime may alter during the cardiac cycle so estimates of mean velocity could be erroneous.
- An estimate of the mean cross-sectional area and mean velocity over a cardiac cycle will not give a true measurement of mean volume flow-rate because the two are not linearly related. Attempts at instantaneous measurement of both mean velocity and mean cross-sectional area are difficult to coordinate due to signal processing limitations.

Many modern duplex scanners do have algorithms for calculating blood volume flow rate and reasonable estimates can be achieved on vessels between 4 and 8 mm in diameter (Evans *et al* 1989).

Alternatively, a number of estimators have become popular to quantify flow in a way suitable to gain clinical information. The measurement of the maximum frequency shift is relatively straightforward and can provide a useful insight into flow abnormalities. Figure 7.14 shows the type of variation that can occur over a cardiac cycle albeit for forward flow only. Mo *et al* (1988) compared different methods of estimating the maximum frequency.

Although 'waterfall'-type displays are sometimes used in research most modern Doppler blood-flow analysers display the FFT frequency spectrum of figure 7.9 as a sequence of vertically orientated frames. The location of a sample frame is shown in figure 7.14. These displays are produced in

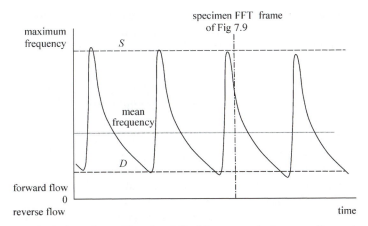

Figure 7.14. Typical maximum Doppler-shifted frequency during a cardiac cycle.

rolling format on the monitor and are referred to as *sonograms*. The intensity information is in the *z*-axis direction (out of the page) and is shown as a colour code by this type of analyser.

Understanding the data gathered becomes an issue of pattern recognition. Various algorithms have been invented over the years in an attempt to automate the feature extraction process. Quantifying parameters in use are:

- the ratio S/D,
- the pulsatility index,

$$\frac{S - D}{\text{mean velocity}} \qquad (7.53)$$

- Pourcelot's resistance index,

$$\frac{S - D}{S}. \qquad (7.54)$$

To obtain a value for S some threshold has to be chosen above background. Also, low frequency filters are used so that values of D are not influenced by extraneous vibrations. The mean velocity is estimated over the cardiac cycle. This can be conveniently done as a rolling average using the mean frequency from a FFT frame.

Conclusion

Although the measurement of flow has been carried out for millennia it still attracts much research. This chapter has outlined some of the applications of signal processing to the more challenging flow fields and illustrates the expertise needed across the range of engineering physics to produce a working device.

References

Angelsen A J B 1980 'A theoretical study of the scattering of ultrasound from blood' *IEEE Trans. Biomed. Engineering* **BME-27**(2) February

Beck M S and Plaskowski A 1987 *Cross-correlation Flowmeters* (Adam Hilger)

Bentley J P 1995 *Principles of Measurement Systems* 3rd edition (Longman)

Brody W R 1974 'Theoretical analysis of the CW Doppler ultrasonic flowmeter' *IEEE Trans. Biomed. Engineering* **BME-21**(3) May

Burns P N, Halliwell M, Wells P N T and Webb A J 1982 'Ultrasonic Doppler studies of the breast' *Ultrasound Med. Biol.* **8**(2) 127–143

Cousins T 1978 *The Doppler Ultrasonic Flowmeter: Flow Measurement of Fluids* (Amsterdam: North-Holland)

Dickey R K 1975 'Outputs of op-amps have fixed phase-difference' *Designer's Casebook, Electronics*, pp 82–83, 21 August

Evans D H, McDicken W N, Skidmore R and Woodcock J P 1989 *Doppler Ultrasound Physics: Instrumentation and Clinical Applications* (Wiley)

Fox R W, McDonald A T and Pritchard P J 2004 *Introduction to Fluid Mechanics* 6th edition (Wiley)

Gessner U 1969 'The performance of the ultrasonic flowmeter in complex velocity profiles' *IEEE Trans. Biomed. Engineering* **BME-16**(2)

Horowitz P and Hill W 1989 *The Art of Electronics* 2nd edition (Cambridge University Press)

Khandpur R S 1987 *Handbook of Biomedical Instrumentation* (New Delhi: McGraw-Hill)

Mo L Y L, Yun L C M and Cobbold R S C 1988 'Comparison of four digital maximum frequency estimators for Doppler ultrasound' *Ultrasound Med. Biol.* **14**(5) 355–363

Namekawa K, Kasai C, Tsukamoto M and Koyano A 1982 'Real-time blood flow imaging system utilizing autocorrelation techniques' *Ultrasound '82*, ed Lerski and Morley (New York: Pergamon)

Nippa J H, Hokanson D E, Lee D R, Sumner D S and Strandness D E jr. 1975 'Phase rotation for separating forward and reverse blood velocity signals' *IEEE Trans. Sonics and Ultrasonics* **SU-22**(5) September

O'Donnell M, Busse L J and Miller J G 1981 *Piezoelectric Transducers Methods of Experimental Physics* vol 19 (Academic Press)

Rabiner L R and Gold B 1975 *Theory and Application of Digital Signal Processing* (Prentice-Hall)

Sanderson M L and Hemp J 1981 'Ultrasonic flowmeters—a review of the state of the art' *Int. Conf. Advances in Flow Measurement, Warwick University*, 9–11 September, sponsored by BHRA Fluid Engineering

Scott C 1984 'Sounding out ultrasonic flowmeters' *Control and Instrumentation* August

Wells P N T, Halliwell M, Skidmore R, Webb A J and Woodcock J P 1977 'Tumour detection by ultrasonic Doppler blood-flow signals' *Ultrasonics* September

Womersley J R 1957 'An elastic tube theory of pulse transmission and oscillatory flow in mammalian arteries' Wright Air Development Centre, *WADC Technical Report*, TR 56–614

Exercises

1. When air is drawn into the lungs the pressure inside the trachea is found to be about 2.5 Pa below atmospheric pressure. Given that the density of air is about $1.25 \, \text{kg} \, \text{m}^{-3}$ estimate the velocity of air in the trachea. The trachea divides into two, one to each lung. If each branch has 0.65 the cross-sectional area of the trachea what is the velocity of air in a branch? (Assume that the tube walls are rigid and that air is incompressible under these conditions.)

2. A horizontal pipe has a cross-sectional area of A_1 and the liquid in this section flows at a velocity v_1. If the pipe reduces in diameter to A_2 with a corresponding increase in flow velocity to v_2, assuming ideal conditions and that the liquid is incompressible show that

$$v_1 = \sqrt{\frac{2\Delta P}{\rho[(A_1/A_2)^2 - 1]}}$$

where ΔP is the pressure difference between the two sections of pipe. Hence show that the volume flow rate Q is given by

$$Q = A_2 \sqrt{\frac{2\Delta P}{\rho[1 - (A_2/A_1)^2]}}.$$

3. Given that for laminar flow

$$u = U_{\text{max}}\left[1 - \left(\frac{r}{R}\right)^2\right]$$

show that $\bar{u} = \frac{1}{2}U_{\text{max}}$.

4. If the number of clock cycles necessary to reach a correlation in a cross-correlation flowmeter is q, show that

$$\frac{\Delta u}{u} = \pm\frac{1}{2q}.$$

5. In a certain correlation flowmeter it is found that the maximum possible separation of the sensors without losing the correlation is 0.45 m. If a resolution of 2% is required in the velocity measurement, determine
   (i) the clock period if the maximum velocity to be measured is about $0.35 \, \text{m} \, \text{s}^{-1}$.
   (ii) the lowest velocity that can be measured given that the memory buffers are all 256 locations long.

Answers to exercises

Chapter 1

1. Refer to pages 1, 2.

2. See table 1.4.

3. $5.622\,\Omega$; $s = \pm0.006\,\Omega$; $s_m = \pm0.002\,\Omega$
   The standard deviation in the mean should be quoted with R because this indicates the reliability of the estimate of the mean.

4. $\pm2.5\%$.

5. $\pm1.5\%$.

6. $\pm1.8\%$.
   Because the errors given are not standard deviations the formal derivation of the overall error is not applicable. Consequently, the result obtained for the overall error cannot be used to estimate the probability of a result falling within some particular band.

8. (i) See page 11.
   (ii) *Across* is potential difference and *through* is current.
   (iii) *Across* is the temperature difference and *through* is the thermal power.
   (iv) Across is the excess temperature of the tank compared to the sensing element and *through* is the thermal power flowing from tank to sensor.

9. (i) $y_1 = z_{11}\dot{x}_1 + z_{12}\dot{x}_2$, $y_2 = z_{21}\dot{x}_1 + z_{22}\dot{x}_2$

   (ii) $\begin{bmatrix} y_1 \\ y_2 \end{bmatrix} = z\begin{bmatrix} \dot{x}_1 \\ \dot{x}_2 \end{bmatrix} = \begin{bmatrix} z_{11} & z_{12} \\ z_{21} & z_{22} \end{bmatrix}\begin{bmatrix} \dot{x}_1 \\ \dot{x}_2 \end{bmatrix}$

10 (i) If the hot liquid is of small thermal capacity so that its supply of thermal power is limited then the insertion of the sensor would cause the across variable (temperature) to be reduced.

Chapter 2

1. (i) See page 23.
   (ii) See page 24.
   (iii) See page 31.

2. See pages 31, 32.

3. See J P Bentley 1995 *Principles of Measurement Systems* 3rd edition (Longman).

4. See section 2.2.3.

5. (i) $\omega_n = 2.988 \times 10^3 \, \text{rad s}^{-1}$; $\xi = 1.106 \times 10^{-3}$; for light damping $Q \cong (1/2\xi)$ giving $Q = 452$.
   (ii) Sensitivity is defined as d(output)/d(input). In this example the output is ω_n and the input is the density ρ. Differentiating gives

$$\frac{d\omega_n}{d\rho} = -\frac{\omega_n^3}{2\alpha + \beta\omega_n}$$

6. $\Delta f = 5.25 \, \text{Hz}$.

Chapter 3

1. and 2. See the early pages in the chapter.

3. (i) See section 3.2.4.
   (ii) See section 3.2.4 and question 2 in chapter 2.

4. (i) Large pressures and contamination.
   (ii) The radial stress is zero when $r = \sqrt{3}R_o$. The tangential stress is zero at the rim.

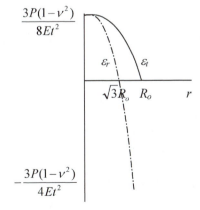

(iii) See figure 3.4.

5. (a) See pages 56 and 60.
    (b) The resolution of the surface mass measurement μ is $3.577 \times 10^{-10}\,\mathrm{kg\,m^{-2}}$.

6. See pages 62 and 63.

7. See page 68.

8. (a) See page 82.
    Low noise but nonlinear response (logarithmic) and poor frequency response.
    (b) (i) From equation 3.74

    $$\eta = \frac{I_0 - I}{I_0} = \frac{I_0 - I_0\exp(-\mu x)}{I_0} = 1 - \exp(-\mu x)$$

    (ii) $i_{\mathrm{p}} = 1.841 \times 10^{-10}\,\mathrm{A}$.

9. See pages 93–96.

10. (ii) The bridge circuit forms two potential dividers so

    $$E_{\mathrm{Th}} = V_{\mathrm{s}}\left(\frac{R_1}{R_1 + R_4} - \frac{R_2}{R_2 + R_3}\right)$$

    Making $R_1 = R_3 = R_0 + \delta R$ and $R_2 = R_4 = R_0 - \delta R$ gives

    $$E_{\mathrm{Th}} = \frac{\delta R}{R_0}V_{\mathrm{s}}$$

    (iii). Approximately $2\,\mu\mathrm{T}$ using a power supply for the bridge of 5.0 V.

Chapter 4

1. See pages 112 and 113 and Table 4.1.

2. See pages 120 and 121.

3. (a) See section 4.2.2.
    (b) There are two ways this could be approached. Either, use a set of source-detector pairs for specific wavelengths over the desired range or, use a white light source and wide-band receiver(s). With the white light source a dispersive element such as a prism or diffraction grating could be used to derive specific wavelengths. For referencing, the source and detector are calibrated for zero attenuation before the fibre is placed in the beam path.

4. (a) See section 4.3.6.
    (b) 143.9; 124.4.

(c) Either, the system can be calibrated for one particular application or, a smart sensor devised with a look-up table for the corrected values.

5. See section 4.3.1. Temperature measurement could be achieved using an expansion element fastened to the reflector.

6. See section 4.3.3. Since the absorption is logarithmic a sensor with a logarithmic response would be appropriate such as the photodiode used in photovoltaic mode.

7. From equations 4.84 and 4.74, $E = E_0 \exp(-x/\chi)$ where

$$\chi = \frac{\lambda_0}{2\pi(n_1^2 \sin^2 \alpha - n_2^2)^{1/2}}.$$

Now intensity goes as E^2 so 4.84 becomes, $E^2 = E_0^2 \exp(-2x/\chi) = I$. So the sensitivity is given by,

$$\frac{dI}{dx} = -\frac{2}{\chi}I.$$

8. See pages 151–153. The major difficulty is the nonlinearity of the Kerr effect.

Chapter 5

1. 8-bit ADC; $>3 \times 10^3$ bit s^{-1} because of the Shannon–Nyquist sampling theorem.

2. (i) Feedback. (High speed, high resolution.)
   (ii) Dual slope. (Can be configured to reject power-line interference.)
   (iii) Feedback. (Moderate frequency response and resolution available at a low price).

3. 2.304×10^7 bits s^{-1}.

4. 6-bit; 3.932×10^7 bits s^{-1}; flash ADC.

5. A 10-bit ADC is needed. See pages 191–193 for useful features.

6. See section 5.6.1. The logic statements can be arranged in a number of ways and are relatively straightforward. The difficulty is in imagining all possible scenarios and writing in the appropriate machine response.

7. See section 5.6.2. Value judgements have to be made as to what the *about* weightings should be when creating the membership functions for the fuzzy sets.

8. See section 5.6.3.

9. See section 5.6.3.

Chapter 6

1. See pages 219, 220.

2.   (i) The heating of the water in the tank is a stochastic process. The data ensemble from the thermometers is of a non-stationary, random process. Depending on the exact conditions it may be possible to consider the ensemble as a stationary random process multiplied by a deterministic factor; see equation (6.15).
     (ii) With perfect insulation and no heating the thermometers would all read the same all of the time. The data ensemble would then be of a deterministic, non-periodic process.

3. 1.414 V; 1.155 V; 1.265 V.

4. $\dfrac{1}{V_0}$; 0, $\dfrac{V_0}{\sqrt{6}}$; $\frac{1}{8}$.

5. $R(\tau) = \dfrac{16}{\pi}\,\text{sinc}(200\tau)$.

6. An auto-correlation of the signal from the vibration sensor will separate the noise from the periodic function. The amplitude of the periodic signal may be determined from the amplitude of the latter part of the autocorrelogram, see pages 245–247.

7. $R_{vv}(\tau) = 0.005\cos 100\tau + \text{sinc}\,200\tau$.

8. The screen cable is needed to exclude interference. However, because of the relatively weak signals in the system variation in local earth potentials at the mains frequency can inject *hum* into the output. This is because the difference in earth potentials at the two ends of the cable cause current to flow in the screen. This results in an additional signal voltage at the receiver. Using twin-core, co-axial cable means that the screen does not need to be used for the signal.

9. 0.103 K.

10. See pages 239–244.

11. See section 6.5.2.

12.   (i) The unwanted random signals may be due to intermittent arcing of the electrical contacts of the tram, vibrations in the ground from sources other than the tram, noise within the seismometer circuitry or cabling, electrical noise in the ground itself, or other passing vehicles.
         The unwanted deterministic signals may be due to electromagnetic waves from the pantograph or the electrical machinery or electronics on the tram. Extraneous radio waves may be present from mobile phones

etc. The power-line return for the tram is through the track and will affect the local earth. This could appear in the screen of the single-core, co-axial cable.

(ii) See sections 6.4.2 and 6.5.2.

Chapter 7

1. $2.0\,\mathrm{m\,s^{-1}}$; $1.54\,\mathrm{m\,s^{-1}}$.

2. For horizontal, ideal flow Bernoulli's equation reduces to,

$$P_1 + \tfrac{1}{2}\rho v_1^2 = \text{constant}$$

So, for the two points along a streamline,

$$P_1 + \tfrac{1}{2}\rho v_1^2 = P_2 + \tfrac{1}{2}\rho v_2^2$$

Re-arranging,

$$\tfrac{1}{2}\rho v_1^2 = P_2 - P_1 + \tfrac{1}{2}\rho v_2^2 = -\Delta P + \tfrac{1}{2}\rho v_2^2$$

ΔP is negative because, from Bernoulli's equation, if $v_2 > v_1$ then $P_2 < P_1$. The equation of continuity can be expressed as, $\rho_1 A_1 v_1 = \rho_2 A_2 v_2$ and since the liquid is incompressible this reduces to, $A_1 v_1 = A_2 v_2$. Substituting for v_2 above gives

$$\tfrac{1}{2}\rho v_1^2 = -\Delta P + \tfrac{1}{2}\rho \left(\frac{A_1}{A_2} v_1\right)^2$$

Re-arranging for v_1 yields,

$$v_1 = \sqrt{\frac{-2\Delta P}{\rho\left[1 - \left(\frac{A_1}{A_2}\right)^2\right]}} = \sqrt{\frac{2\Delta P}{\rho\left[\left(\frac{A_1}{A_2}\right)^2 - 1\right]}}$$

Finally, given that $Q = A_1 v_1$,

$$Q = A_1 \sqrt{\frac{2\Delta P}{\rho\left[\left(\frac{A_1}{A_2}\right)^2 - 1\right]}} = A_2 \sqrt{\frac{2\Delta P}{\rho\left[1 - \left(\frac{A_2}{A_1}\right)^2\right]}}$$

3. Using equation 7.19,

$$Q = \int_0^R u(r)2\pi r\,dr = \int_0^R U_{\max}\left[1 - \left(\frac{r}{R}\right)^2\right]2\pi r\,dr$$

$$Q = 2\pi U_{\max}\int_0^R \left[1 - \left(\frac{r}{R}\right)^2\right]r\,dr = 2\pi U_{\max}\int_0^R \left[r - \frac{r^3}{R^2}\right]dr$$

$$Q = 2\pi U_{max} \left[\frac{r^2}{2} - \frac{r^4}{4R^2}\right]_0^R = 2\pi U_{max} \left[\frac{R^2}{2} - \frac{R^4}{4R^2}\right] = \pi U_{max} \frac{R^2}{2} = \tfrac{1}{2} A U_{max}$$

But by definition $Q = A\bar{u}$, so $\bar{u} = \tfrac{1}{2} U_{max}$.

4. Use equation 7.27.

5. (i) 50 m s.
    (ii) 0.035 m s^{-1}.

Index